EARLY YEARS
ACTIVITY CHEST

ICT
activities

British Library Cataloguing-in-Publication Data
A catalogue record for this book is available from the British Library.

ISBN 0 439 01733 5

DEDICATION
This book is dedicated to my parents who gave me, my brother James, and so many others, such a memorable childhood. Special thanks to our mum, Gwen, who died in March 2001. She brought me up to believe that I could do anything if I tried. I owe her so much.

ACKNOWLEDGEMENTS
The publishers gratefully acknowledge permission to reproduce the following copyright material:

Jillian Harker for 'Pass the message, please' and 'Mr Teazle's toyshop' © 2001, Jillian Harker, both previously unpublished; **Brenda Williams** for 'Children come from many lands' and 'Where we live...' © 2001, Brenda Williams, both previously unpublished.

Every effort has been made to trace copyright holders and the publishers apologize for any inadvertent omissions.

AUTHOR
Anne Farr

EDITOR
Jane Bishop

ASSISTANT EDITOR
Lesley Sudlow

SERIES DESIGNER
Lynne Joesbury

DESIGNER
Nari Sandhu

ILLUSTRATIONS
Debbie Clark

COVER PHOTOGRAPH
Martyn Chillmaid

Text © 2001 Anne Farr
© 2001 Scholastic Ltd
Designed using Adobe Pagemaker
Published by Scholastic Ltd, Villiers House,
Clarendon Avenue, Leamington Spa, Warwickshire CV32 5PR
Printed by Bell and Bain Ltd, Glasgow
Visit our website at www.scholastic.co.uk

4567890   567890

CONTENTS

CONTENTS

Introduction

Today, information and communication technology is part of our everyday life and will become more so in future years. We have become dependent on 'the hole in the wall' to obtain cash; many adults and children have regular access to computer systems; we send e-mails and faxes and use mobile phones for easier contact with family, friends and business. Much of the familiar equipment that we have in our homes, for example, toasters, microwaves, washing machines, security systems, televisions and videos, is controlled by information and communication technology.

Young children are surrounded by ICT at home, in their educational setting and in everyday life. They see people working electronic tills and bar-code readers in supermarkets. They understand that washing machines have different settings to wash, rinse and even, in some cases, dry our clothes. Some children may be able to record their favourite television programmes, inserting the videotape and setting the time and channel more confidently than many adults!

Watching children of two, three and four years of age manipulate electronic game controls with great dexterity is nothing short of awe-inspiring. Similarly, the hand–eye co-ordination displayed by these children using a computer 'mouse' or tracker ball to move around or navigate their 'programs' on the computer is a sight to behold. Ask children to direct you how to do it and you will no doubt agree that children's attitudes and responses to technology are amazing.

We are all aware of the knowledge, skills and understanding that young children have of information and communications technology and the wealth of experiences that some children bring from home. Being able to choose, use and apply ICT is an important skill that children will certainly need in their future lives. However, we must remember that ICT is not just about computers, but about a whole range of other things such as cookers, microwaves, fax and photocopying machines, telephones, cassette recorders, electronic keyboards, digital cameras, programmable toys, electronic scales and calculators. To support children in becoming competent and confident users of ICT, we need to integrate these commonplace items into the early years curriculum.

The aim of this book is to provide educators and carers of three- to five-year-olds with a range of practical dip-in activities and supporting photocopiable sheets to help teach ICT awareness and to develop technological understanding and skills. These activities are delivered through the six areas of the Foundation Stage, although many will address several areas. The activities involve children in a variety of experiences which encourage and develop listening, manipulating, speaking, observing, communicating, problem solving, numeral recognition, mathematical skills, pre-writing skills and writing skills. From compiling a cassette of favourite jokes to sending a fax, all the activities in this book have been carefully planned to extend children's learning.

How to use this book

There are six chapters, each covering one of the six Areas of Learning identified by the QCA in its document *Curriculum guidance for the foundation stage*. The ideas suggested can be applied equally well to the documents on pre-school education published for Scotland, Wales and Northern Ireland. Each chapter contains eight activities which cover a wide spectrum of ideas and provide a well-balanced curriculum.

The activities featured in the book follow an identical format with the sections headed 'Learning objective', 'Group size', 'Timing', 'What you need' and 'What to do'. The sections on 'Support' and 'Extension' at the end of each activity offer practical suggestions for adapting or extending the main activity to suit younger or less able children, and older or more able children respectively.

In addition, the 'Home links' section offers initiatives to help establish a sound link between the home and the school or nursery. 'Multicultural links' are also suggested where possible to help the children gain or maintain an awareness of other countries and cultures, festivals and

traditions. From listing the resources needed for each activity to suggestions for work with a variety of different groups of children, this book covers the key elements of effective lesson planning.

Using the photocopiable sheets

There are 24 photocopiable sheets in this book, which aim to support or extend individual activities. The photocopiable sheets use a variety of approaches to consolidate and strengthen children's previous experiences and understanding of ICT and stimulate their interest in technology, for example, children can write a message on the photocopiable fax sheet on page 65 and then send it.

Stories and poems have been written specially to use with some of the activities such as 'Pass the message, please' on page 57 and 'Children come from many lands' on page 59.

Many of the photocopiable sheets can also be used as 'stand alone' activity sheets to help develop or reinforce particular skills.

Resources

Pre-school children learn best through hands-on activities – talking, active participation, repetition, asking questions and solving problems – all of which require concrete experiences. Children need concrete experiences in order to manipulate and work the objects at their disposal, and in so doing they learn to make sense of the world. They enjoy the challenge of using different resources, and by providing a variety of ICT experiences, you can help to stimulate their interest and enthusiasm.

The activities in this book make use of a wide variety of different resources, which are available in many early years environments, or can be easily obtained. Resources used for activities in this book include cassette recorders and tapes, cookers, calculators, computer systems, printers, CD-ROMs, digital and ordinary cameras, mobile and landline phones, fax machines and other everyday technology.

Links with home

The connection between home and the early years group is a vital one and parents and carers need to be actively involved in ensuring the continuity of children's education. To strengthen this link, each of the activities contains a practical suggestion that can be followed through at home. Invite parents and carers to help their children to complete these small activities, and use the ideas to demonstrate to parents and carers how much fun can be gained from ICT. Above all aim to encourage a positive attitude towards the subject.

Parents and carers often like to know more about the projects that their children are doing, so good communication is vital. Most of the photocopiable sheets within this book could be used as 'Home links' sheets to help reinforce the work that you have introduced. In addition, photographs of the children taken while they work and play, and displayed on a prominent notice-board, usually generate lots of interest.

Multicultural links

Many of the activites in this book provide a multicultural activity idea, which will help children to recognize that there are many different cultures and religions in the world around us, and that they all have different celebrations and festivals. The main aim is to encourage the children to respect, be tolerant of and be sensitive to the individual differences and cultural and religious backgrounds of others. Encourage opportunities for the children to speak in their mother tongue and to share their favourite stories, songs, recipes and so on. There is so much that we can learn from other cultures.

Opportunities to use ICT

It is important for children to realize that ICT is not just about computers but that it focuses on a wide range of items used in the everyday world. A variety of ongoing, informal activities can be used to promote an understanding and interest in technology, for example, by setting up imaginative play situations such as a garage, café and a supermarket, and including items such as telephones and cash tills. Children need to be aware of how technology is used in everyday life, for example, a washing machine – programmed to wash the clothes; a cooker – to roast, grill and boil food; a video-recorder – to record favourite programmes on television; faxes, phones and e-mail – to communicate with other people. In conclusion, support children's desire to learn more about ICT in everyday situations and help them to find pleasure in what they do. Above all, encourage the children to have fun.

CHAPTER 1

Personal, social and emotional development

This chapter contains activities to encourage personal, social and emotional development. These include baking cakes to inspire children to take turns while working as part of a group and using a camera to take photographs, then looking at facial expressions to recognize why we show different feelings.

GROUP SIZE
Any size.

TIMING
Ten to 15 minutes.

HOME LINKS
Invite the children to share their favourite pieces of music with their families. Make a class 'music library' on tapes for the children to borrow.

MULTICULTURAL LINKS
Provide a wide selection of music from different cultures and add these to the class 'music library'.

MUSIC TIME

Learning objective
To maintain attention, concentrate, and sit quietly when appropriate.

What you need
Coloured card; felt-tipped pens; a suitable area to set up as a 'listening area'; tape recorder or CD player; headphones; blank tapes; a selection of tapes or CDs, for example, 'The Four Seasons' by Vivaldi from *Hall of Fame 2000* (Classic FM); 'Circle of Life' from *Pan Pipe Moods* (Polygram); and 'Babar the Elephant' by Julian Lloyd Webber from *Cradle Song* (Phillips).

Preparation
Choose three or four pieces of music. If using CDs, it might be useful to record these onto a tape. Set up the 'listening area' with the tape recorder, headphones and tapes. Cut up the card and write the name of each selected piece of music onto several cards.

What to do
Gather the children together. Hold up a tape and ask, 'Do you know what this is?', 'How is it used?', 'What might you hear on a tape?'. Give further examples such as music, poems, songs, rhymes, stories, people talking and different sounds. Demonstrate how to use the tape recorder. Encourage the children to listen carefully. Ask, 'What can you hear?', 'Is it loud or soft?'. Invite the children to decide which is their own favourite piece of music and then to find its title on one of the cards.

Organize the children into small groups, then one group at a time, invite the children to the 'listening area' to choose their favourite piece of music and to collect a card with its title.

Next, ask the children to sort the cards to find the most popular choice. Select a child to operate the tape recorder to play the chosen piece of music once again.

Support
Limit the number of selections to no more than three for younger children. Include favourite songs and nursery rhymes.

Extension
Encourage older children to listen to a range of music and select their favourites. Add these to the class collection with their titles on cards.

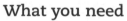

CAKES FOR TEA

Learning objective
To work as part of a group, taking turns.

What you need
Cooker; electronic scales; simple cake recipe and required ingredients; baking tins; bowls.

Preparation
Word-process the recipe. Use large print and an appropriate font. Print out the recipe, make an A3 size copy and display it. Prepare the cooking area. Organize the ingredients and assemble all the resources together. (NB Check for any food allergies or dietary requirements.)

What to do
Explain to the children that you are going to make cakes for them to share with their friends. Ask the children to name some of their favourite cakes such as muffins, fairy cakes, chocolate brownies, carrot cake. Invite the children to suggest what they will need to make some cakes, for example, cooker, ingredients, bowls, spoons and so on. Show the children the recipe and read it aloud, pointing out the ingredients. Explain that there is a special time and oven temperature for the cakes to cook. Ask, 'What would happen if the cakes were left in the oven too long?', 'What would the cakes be like if they were not left in the oven long enough?'.

Invite the children to look at the cooker. Point out where the temperature control is for the oven and also the timer. Help the children to set the required temperature. Point out health and safety issues.

Discuss with the children the ingredients that are needed to make the cakes. Ask, 'How much of each ingredient do we need?', 'How do we work out the correct amount?'. Demonstrate how to use the scales. Involve the children in weighing out some of the ingredients. Encourage them to take turns in preparing the cakes. Set the oven timer and remind the children to listen for it ringing. When the cakes are cooked leave to cool.

Finally, gather the children together and share out the cakes.

Support
Help younger children to weigh out the ingredients. Discuss the process and encourage turn-taking.

Extension
Encourage older children to read the recipe and allow them to do as much as possible themselves.

GROUP SIZE
Two to four children.

TIMING
Ten minutes.

HOME LINKS
Ask parents and carers to involve their children in making and receiving telephone calls, allowing them to help dial telephone numbers and answer the telephone at home.

MULTICULTURAL LINKS
Encourage the children to speak in their home language to make and receive telephone calls.

HELLO – IT'S ME!

Learning objective
To be confident to try new activities.

What you need
A selection of telephones, including dial, press-button and mobile phones; tape recorder; blank tape; the photocopiable sheet on page 62.

Preparation
Make a tape-recording of different telephones ringing and an answering machine taking a call. Assemble the phones together on a table-top.

What to do
Gather the children together and show them the different telephones. Encourage the children to think about how they are different. Ask, 'Do they all have a keypad?', 'Which part of the telephone do you use for talking?', 'Which part is used for listening?'. Explain that the telephone is an important way of telling people things and finding out information. Ask, 'What do we need to do to make a telephone call?', 'How do we know when someone is trying to ring us?'. Talk about making and receiving phone calls and how we use the phone for different things, for example, making appointments, speaking to friends and family, calling emergency services and asking for help.

Ask the children to read the numbers on the telephones and explain how to make a call. Invite individuals to 'dial' a number and pretend to speak to someone on the phone. Play the tape of the different telephones ringing. Choose someone to answer the call. Ask, 'What sometimes happens if no one is able to answer the telephone?'. Play the tape-recording of the answering machine. Give each child a copy of the photocopiable sheet and discuss who is making the telephone calls and what they might be saying, and invite the children to draw pictures in the spaces.

Place the phones in the role-play area with the tape recorder and invite the children to use the phones to make and answer calls.

Support
Ask an adult to 'ring' younger children. Encourage the children to answer the call and make pretend telephone calls to their friends and family.

Extension
Ask older children to make a telephone call independently. Encourage them to dial their own telephone number if they know it. Make sure that they can also answer the telephone.

GROUP SIZE
Small groups.

TIMING
Ten to 15 minutes.

HOME LINKS
Give each child a copy of the photocopiable sheet on page 63 to draw smiles on the faces. Invite the children to collect pictures from magazines of other people smiling for the class display.

SMILE, PLEASE!

Learning objective
To show a range of feelings when appropriate.

What you need
Camera; film; access to photocopier; mirrors; paper for mounting; display space; Blu-Tack; photocopiable sheet on page 63.

Preparation
In advance, take each child's photograph and one of the group together. Arrange for the film to be developed. Set up an area to display the photographs and make a caption saying 'Smile, please!'.

What to do
Gather the children together and smile at them. Look around the group and comment on their lovely smiles. Remind the children of the time that you took the photographs of them. Explain that the film in the camera has been developed into prints. Show the children the photographs and ask them to identify each child. Ask, 'How can we tell who it is?'. Cover the top, bottom, left and right sides of the photographs in turn. Can the children still tell who it is? Point out that the children are smiling in different ways, for example, with teeth showing or with lips together.

Look at the group photograph, Ask, 'Is everybody smiling?', 'Why do we ask people to smile when we take photographs?', 'Do our faces look different when we smile?', 'How do we feel when people smile at us?', 'If people are smiling are they usually happy or sad?'.

Work with two or three children at a time to photocopy their own photograph. Give out the mirrors and encourage the children to look at themselves smiling. Suggest that they make different faces, for example, a sad face, a happy face and an angry face. Give each child a copy of their own photograph to take home and share with their family. Display the group and individual photographs and encourage the children to write their own captions.

Support
Younger children might need a helper to encourage them to make different faces in the mirror.

Extension
Invite older children to draw pictures of their friends smiling and add these to the display.

GROUP SIZE
Two or three
children.

TIMING
Ten to 15 minutes.

PLAY IT AGAIN

Learning objective
To select and use resources independently.

What you need
Tape recorder; headphones; selection of nursery rhymes, stories and songs on tape; table and chairs; plastic containers to hold tapes; sticky labels; felt-tipped pens.

Preparation
In a suitable area of the room set up the tape recorder and headphones on a table. Write individual labels such as 'nursery rhymes', 'songs', 'music' and 'stories' and attach to the containers. Sort the selection of tapes into the appropriate containers. Pick out a favourite nursery rhyme or song.

What to do
Gather the children together. Point out the 'listening area' and explain how the tapes are organized. Tell the children that you have selected a favourite song and invite them to listen to it. Demonstrate how to use the start, stop, rewind and fast forward buttons on the tape recorder and then play the song or nursery rhyme. Invite the children to tell you some of their favourite stories and songs. Ask, 'What is it that makes this special for you?'. Show the children where to look for the content of the tapes. Explain that they can listen to the tapes to find their favourites. Ask, 'How do you rewind the tape to listen again?', 'How do you start and stop the tape recorder?'.

Invite the children to use the 'listening area'. Make sure that the children can operate the tape recorder independently. Encourage them to take turns choosing the tapes and operating the tape recorder to find their favourite pieces. Suggest that the children discuss their favourites with the others. After the children have finished, ask them to come and tell you some of the things that they enjoyed and to show you which tapes they have used. Let different children operate the tape recorder. Invite a different child each day to choose a piece of music to play to the rest of the group at a special time.

Support
Stick different-coloured labels on the start and stop buttons to make it easier for younger children to operate the tape recorder.

Extension
Encourage older children to pick out their favourite pieces by fast-forwarding and rewinding the tapes. Explain how to use the number counter to select precise parts of the tape.

HOME LINKS
Ask the children to think about how they select TV programmes, computer games or videos at home.

GROUP SIZE
Two or three
children.

TIMING
Ten to 15 minutes.

WHAT'S THAT?

Learning objective
To be interested and excited to learn.

What you need
Computer system with CD-ROM drive; selection of CD-ROMs, for example, *Winnie the Pooh Story Book* (Disney), *Dr Seuss's ABC* (Broderbund Living Books); *Tortoise and the Hare* (Broderbund Living Books); *Just Grandma and Me* (Broderbund Living Books) or *Little Monster at School* (Broderbund Living Books).

Preparation
Load the CD-ROM and familiarize yourself with the program. Find out how to navigate through the program, select options and move forwards or backwards on screen. Explore what happens if you point and click on various objects. If using adult support, make sure that they also understand what to do.

What to do
Gather the children around the computer. Show them the title page and explain that they are going to work together to explore the story. Demonstrate how to 'point and click' using the mouse and how to move forwards, backwards and return to the main menu. Point to and read any text on each page and explore some of the interactive features. Ask, 'Did you think that was going to happen?', 'What do you think will happen next?'. Use the illustrations as discussion points, for example, ask, 'Why did the hare decide to have a rest? (*Tortoise and the Hare*), 'Was it healthy for Little Critter to pick the hotdog up off the sand and put it back in the bread roll?' (*Just Grandma and Me*).

Explain to the children that they are going to work together on the computer, taking turns and helping each other to find out what happens. Initially, support them in controlling, pointing and clicking the mouse and then encourage them to explore independently. Invite them to tell you and the other children what they have discovered and to show some of their finds.

HOME LINKS
Invite parents and carers in to share these CD-ROMs with their children.

Support
This activity is particularly suitable for younger children who need to develop their ICT skills and understanding. Use adult support to help the children explore. Encourage younger children to describe what they have found and demonstrate how they found it.

Extension
Suggest older children support others to navigate the CD-ROM. Invite them to demonstrate their favourite interactions.

THE MOUSE RAN UP THE CLOCK

Learning objective
To form good relationships with peers.

What you need
Programmable toy (Pip or Roamer); floor space; large piece of thick card; smaller piece of coloured thick card; paper fastener; thick felt-tipped pens; fabric; sticky labels; dice numbered 1–6 (spotted); dice shaker; adult support.

Preparation
Transform Pip or Roamer into a mouse by adding eyes, ears, whiskers, tail and mouth. Make a grandfather clock by dividing a large piece of thick card into seven squares to form a grid. These should allow Pip or Roamer to fit in comfortably in width and length. Use a thick felt-tipped pen to write individual numerals from 1 to 6 in order in each section and in the seventh section draw a clock-face. Make a pair of card

hands and secure to the clock face with a paper fastener. Place the grid in a large space and put the dice and shaker nearby.

What to do
Gather the children around the grandfather clock and recite the rhyme 'Hickory, Dickory, Dock'. Encourage the children to join in while you point to the clock and the programmable toy. Ask, 'What does the programmable toy look like?', 'What did the mouse do in the nursery rhyme?'. Point out the keypad and remind the children how to operate the programmable toy. Demonstrate the game by rolling the dice. Count the spots and show the children how to program the 'mouse' the appropriate number of moves. Repeat this until the 'mouse' reaches the clock-face. Then program Pip or Roamer to run down the clock.

Invite the children to work in pairs. Stay with the group and encourage them to 'take turns' operating the 'mouse'. Help them to estimate how many steps they will need to reach the top of the clock. Encourage the children to concentrate and watch what the other pair is doing. Step back once the children are confident with the game and each pair are co-operating and supporting each other. This may take several sessions with your encouragement and support.

Support
Younger children will need adult support, However, it is useful for children who find it difficult to work together, or for children who need to gain more confidence in mixing with others.

Extension
Encourage older children to work together to program the 'mouse'.

GROUP SIZE
Any size.

TIMING
15 to 20 minutes.

HOME LINKS
Let the children take home a copy of the poem to share with their families. Encourage parents and carers to look for pictures of children from around the world, and information books about life in other countries. Suggest that they borrow books from the library to share with their children at home.

MULTICULTURAL LINKS
Invite parents and carers to come in to share aspects of their culture with the children. Make a collection of artefacts, including music, and display these in the setting.

ALL ABOUT US

Learning objective
To respect the cultures of other people.

What you need
Photographs, pictures and information books showing children from around the world; computer system; *Powerpoint* program; scanner; artefacts and music from different countries; the photocopiable sheet on page 59.

Preparation
Scan each of the photographs and pictures and save onto a floppy disk. Load the *Powerpoint* program. Insert the pictures to make a 'slide show' and save. Display the information books and artefacts on a table or cupboard near the computer and set up the tape recorder and tape nearby.

What to do
Gather the children together around the computer. Read the poem on the photocopiable sheet on page 59. Explain that there are many people in the world and that we are all different. Ask the children to look at the person sitting next to them. Point out that children can have straight or curly, long or short hair, and that they can also have different coloured eyes, hair and skin. Talk about how children wear different clothes, speak different languages and listen to different types of music (play examples from different countries). Tell the children that although some children may live in different homes and enjoy many kinds of foods, they like to play just as they do.

Explain that you have collected pictures of children from all around the world. Show the children the 'slide show' and encourage discussion. Show the artefacts, explaining what they are, which country they come from and how they are used. Allow the children to touch these where possible. Help the children to operate the 'slide show' independently.

Support
Younger children may need adult support to operate the 'slide show'.

Extension
Encourage older children to recognize that although they are all different, they are all children and do the same sort of things.

CHAPTER 2

In this chapter you will find activities to support children's communication, language and literacy skills. The ideas include looking at a 'living book' on a computer, investigating e-mails, role-playing a secretary's office, sending and receiving fax messages and recording jokes onto a tape recorder.

Communication, language and literacy

GROUP SIZE
Any size for introduction; individuals or pairs with for activity.

TIMING
Ten to 15 minutes.

a b c d e f g h

a b c d e f g h

a b c d e f g h

a b c d e f g h

HOME LINKS
Encourage the children to find examples of different text styles in newspapers, magazines, cards and books at home and to bring them in for the display.

MULTICULTURAL LINKS
Use packages that allow the children to write in their mother tongue.

WRITING FOR FUN

Learning objective
To experiment with text.

What you need
Computer system; word-processing package; colour printer (if available); brightly-coloured A4 paper; display board; blank cards; adult support.

Preparation
Set up the computer and load the word-processing package. In advance, write some text, for example, names, the alphabet, familiar words and phrases. Use different font styles, colours and sizes of text. Print out examples and display on a board.

What to do
Gather the children around the computer. Explain that you are having fun writing messages using a word-processing program. Demonstrate that by pressing keys on the keyboard you can make letters, numbers or symbols appear on the monitor's screen. Next, show how you are able to send messages to the printer to ask it to print your work onto paper.

Point out to the children how you can choose different fonts, large or small print and colours from the menu. Show the examples that you have prepared. Ask, 'Which font style is your favourite?', 'Is large or small text best for making posters or labels?'

Arrange for the children to work on the computer with an adult – either as individuals or in pairs. Ask each child in turn to press any letter or numeral keys on the keyboard and to look for them on the monitor. Encourage them to write their name and then explore the keyboard. Print out their writing. Then support them in changing the style of the font, size and colour of their text and print this out. Give the children the opportunity to 'write' a caption for themselves. Print out three copies of their work – one for display, one for ICT records and one to take home.

Support
Write younger children's names on caption cards for them to copy. Use the computer to support emergent writing.

Extension
Encourage older children to write their captions for display. Ask them to show how they made their font choices from the menus.

GROUP SIZE
Four to six children.

TIMING
Ten to 15 minutes.

STORY TIME

Learning objective
To listen with enjoyment and respond to stories.

What you need
Computer system with CD-ROM drive and speakers; a Broderbund Living Book CD-ROM such as *Just Grandma and Me*, *Green Eggs and Ham*, *Dr Seuss's Cat in the Hat* or *Harry and the Haunted House*; the story-book versions of these titles if available; adult support.

Preparation
Load the CD-ROM and familiarize yourself with the navigation of the book. Make sure that the speakers are set appropriately.

What to do
Gather the children around the computer. Explain that we can read and enjoy stories from books, from living books on the computer and listen to stories on tapes. If the story-book is available, read it as you go through the CD-ROM. Point out different features of the book, for example, the title and the author, look at the illustrations and refer to the page numbers. Start the 'living book'. Choose the option to read the story.

Look at the different features. Ask, 'Are they the same as in the book?', 'What is different?'. Encourage the children to think what happens next in the story. Ask, 'What happened before?'. Encourage the children to discuss their favourite parts and characters.

Invite the children to read the book themselves. Let individuals take turns to move to the next page. Ask an adult to stand close by to observe them and give support where necessary.

At another time, organize the children to 'play along' and interact with the story to see what exciting things happen. Ask, 'How is the living book different from the real book?', 'Which parts did you like best?'.

Support
Help younger children to navigate the story. Read out and point to the words and encourage the children to join in.

Extension
Let older children work on the computer in small groups. They may also like to support younger children.

HOME LINKS
Encourage the children to bring in their favourite stories and story tapes from home to share with the group.

MULTICULTURAL LINKS
Provide a wide selection of story-books and tapes from other cultures for everybody to enjoy. Also include books and tapes in the children's mother tongue.

GROUP SIZE
Six to eight children.

TIMING
Ten to 15 minutes.

HOME LINKS
Ask parents and carers to allow their children to speak to friends and relatives on the telephone.

MULTICULTURAL LINKS
Provide tapes in the children's mother tongue so that they can listen to and speak their own language.

PASS ON THE MESSAGE

Learning objective
To use language to imagine and recreate roles and experiences.

What you need
Landline telephone or mobile phone; notepads; pencils; tape recorder; blank tape; the photocopiable sheet on page 57.

Preparation
Assemble all the objects together and place in the role-play area. Make a tape recording of a phone ringing and add a message to be passed on to another person, for example, 'If you are going to Peter's party tomorrow, can I come with you, please?' or 'Do you want anything from the shops today?'. Make a copy of the photocopiable sheet on page 57. Arrange for someone to call you on the telephone at a given time.

What to do
Set up the telephone or mobile phone together with the tape recorder and tape. Invite the children to sit in a circle, then read the story on the photocopiable sheet. Explain that it is really important to pass on messages to people. Ask, 'What happened in the story?' Tell the children that you are going to play 'Chinese whispers' by giving a message that has to be passed on until everybody has heard it. Whisper a simple phrase such as, 'Have you seen my cat?'. Ask the child sitting next to you to pass on the message until it has gone around the circle.

Invite individual children to say the message that they heard. Discuss how messages sometimes change in passing them on. Ask, 'Why is it important to pass on messages correctly?', 'How else do we send messages?'. (We can send notes, e-mails, drawings, and telephone calls.) Alert the person to make the pre-arranged phone call. Answer the phone and take a message. Repeat the message so that the children can hear.

Say that the clinic rang as the doctor is ill so your appointment is cancelled and you must ring for another appointment. Ask, 'Why is the message important?', 'What might happen if you didn't get the message?', 'What do you need to do?'.

Explain that you have taped some messages on the tape recorder in the role-play area for the children to answer. Invite them to pass the message on to someone else. Remind them to listen carefully to the messages and to pass them on correctly.

Support
Limit the number of younger children in the group to three or four. Start the game with a word or phrase until the children become more confident.

Extension
Develop older children's skills by asking them to jot down the message on the notepad.

GROUP SIZE
Up to six children.

TIMING
15 to 20 minutes.

HOME LINKS
Encourage the children to look out for e-mail addresses given on television. If children have access to e-mail facilities at home, ask parents and carers if they could support them in writing to family and friends.

WRITE IT, READ IT

Learning objective

To understand that print carries meaning and that in English it is read from left to right.

What you need

A computer system with the Internet and e-mail access; a printer.

Preparation

Arrange for a friend to send you an e-mail from a fictional character, for example, Little Red Riding Hood, Goldilocks, Little Bo-Peep, Snow White or the giant. Make sure that the text is large enough for the children to see clearly.

What to do

Remind the children of the 'Chinese whispers' game from the activity on page 19. Invite them to think of other ways that messages can be sent, for example, letters, cards, faxes, notes, in bottles, drawings, by telephone or text messages. Explain that we can send and receive messages through the telephone lines as well as through the post. Gather the children together around the computer and ask if any children have used e-mail at home to send messages to family and friends. 'Log on' to gain Internet access and select the e-mail facility. Point out the icons to the children and look for any new mail in the 'Inbox'. Open the e-mail that you have arranged to be sent from a fictitious character. Show the children the e-mail address, the sender and the message. Read out the message to the group, focusing on reading the text from left to right and from top to bottom, following the print with your finger if necessary. Print out the message. Invite the children to write an e-mail back to the character. Choose the 'write ' or 'reply' option. Ask the children for suggestions to write and act as scribe. Show them how to send the e-mail.

At a later time, ensure that the children receive a reply to their e-mail and have the opportunity to read it. Display the e-mails for everyone to see and read.

Support

Make sure that the message is very simple for younger children and ask an adult to help them to read it.

Extension

Invite older children to work in pairs to send their own e-mails. Ask an adult to help them to write the e-mail address and send it.

IN THE OFFICE

Learning objective
To attempt writing for different purposes such as lists and letters.

What you need
Computer system; word-processing package; printer; A4 paper; role-play area set up as an office with a selection of objects (staplers, hole-punch, pens, notebooks, calendar, diary, paper clips, scissors, desk tidy, erasers, Post-it notes, sticky tape, labels, envelopes, an old typewriter, telephone, writing paper, files and folders); the photocopiable sheet on page 64.

Preparation
Set up an office area in the room. Place the computer system in there with other suitable objects. Load the word-processing package. Make several copies of the photocopiable sheet on page 64 and add 'jobs' for the secretary to do such as ring the library, arrange a visit to the fire station and check the diary for dates. Arrange to take small groups of children to visit the secretary's office.

What to do
After each group has visited the secretary's office, gather the children together. Ask, 'What did you see in the secretary's office?', 'What sort of things does a secretary do?', 'What does she/ he use?', 'Where does he/she keep things?'. Point out the role-play office and ask the children what they would expect to find in there. Show them some of the objects in the office and invite them to describe how they are used. Ask, 'Who works in an office?', 'What things would they need?'. Remind the children how to use the word-processing package, pointing out the 'Spacebar', 'Backspace', 'Enter' and 'Caps Lock' keys.

Organize the children into small groups to play in the office. Encourage them to use the word-processor to write letters, make lists and create labels. Place copies of the photocopiable sheet on page 64 in the office area for the children to role-play. Support the children in printing out their work.

Support
Encourage younger children to play with older children in the office area. Ask an adult to observe the role-play and intervene when appropriate.

Extension
Introduce an event that requires a writing activity for older children, for example, a letter to parents, a list of names or a poster for a jumble sale.

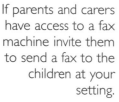

FAX IT

Learning objective
To write their own names and short messages.

What you need
Access to a fax machine; A4 paper; black felt-tipped pens; some examples of faxes; the photocopiable sheet on page 65.

Preparation
Make several copies of the photocopiable sheet on page 65. In advance, arrange to send a fax to a friend or colleague. Ask if they could send a return fax to the children at a pre-arranged time.

What to do
Organize the children into small groups and talk about sending messages to people. Invite the children to suggest different ways of communicating with others, for example, by writing letters, telephoning, sending e-mails, text messaging and using sign language. Explain that many people, especially in offices, use a special item called a fax machine. Show the children some of the examples of faxes pointing out the position of the name, address, date and what the message is about, then read the main part of the message. Invite the children to send a fax message to someone explaining that they are learning about sending messages. Encourage them to think of what to say and act as scribe. On one of the copies of the photocopiable sheet, write a fax message from the group.

Take the children to the fax machine and show them where and how to insert the fax, how to dial the telephone number and how to press the 'Start' button to send the fax. Explain simply that the message goes through the telephone line and reaches a fax machine in another place. The message is transferred onto paper in the other machine and remains there until someone collects it.

Arrange for the children to be at the fax machine when a fax is returned. Listen to the phone line ringing and let the children see how the message comes through the fax machine. Read the message to the children pointing out the main features. Give each child a copy of the photocopiable sheet and let the children make up their own messages for faxing.

Support
Act as scribe for younger children. Write a simple message and include their names.

Extension
Invite older children to fill in all the details on the photocopiable sheet independently and to write a more complex message.

JUST JOKING!

Learning objective
To speak clearly and audibly with confidence and control.

What you need
A tape recorder; blank tapes; a selection of joke books.

Preparation
Before carrying out the activity, show the children how to operate the tape recorder to play, stop, record and rewind.

What to do
Gather the children around and talk about jokes. Ask the children if they know why people tell jokes. Do the children know any jokes? Encourage them to share some of their favourite jokes. Show the children the joke books. Explain that as well as writing down jokes in books, we can record them on tapes for others to share.

Invite the children to record their favourite jokes to make a joke tape. Show the children how to insert the tape and remind them how to use the 'play', 'stop', 'record' and 'rewind' buttons. Explain that they will need to practise saying their jokes before they record them, and then make sure that they speak clearly and loudly into the microphone. Ask the children to practise saying their jokes to each other. Praise their attempts and tell them how well they are doing.

Organize the children into pairs. Ask one child to operate the tape recorder while the other child says his or her joke. Then suggest that each child swaps tasks. When both children have added their jokes ask them to play back the tape. Ask, 'Can you hear each joke?', 'Is it too loud or too soft?', 'Can you repeat the joke?'. Challenge the children to collect more jokes for their tape. Make sure that the jokes are appropriate for others to hear and politically correct!

Support
Help younger children to operate the tape recorder, emphasizing how to play, stop, record and rewind the tape.

Extension
Let older children work together on telling jokes, such as question and answer jokes.

GROUP SIZE
Four to six children.

TIMING
Ten minutes.

HOME LINKS
Give each child a copy of the photocopiable sheet on page 66 to take home. Ask parents and carers to help their children to listen to different sounds made by items found in the home and to list these on the sheet. Invite the children to take a tape recorder and tape home to record their sounds to share with the rest of the group.

SOUNDS AROUND

Learning objective
To explore and experiment with sounds.

What you need
A selection of objects (toaster, electric kettle, food mixer, mobile phone, microwave oven, doorbell, washing machine, radio alarm clock, electric drill); tape recorder; tapes; catalogues and magazines; scissors; adhesive, coloured card.

Preparation
Make a recording of the different sounds that the objects make. Cut out pictures of the same objects from magazines and mount these onto card. Display some of the actual objects on a table.

What to do
Set up the tape recorder and insert the recorded tape. Gather the children together on the carpet. Show them some of the items and invite the children to identify what they are, where they are used and how they are used. Ask, 'Are they used every day, sometimes or not very often?'.

Explain that you have recorded some everyday sounds that can be heard in the home. Hold up some of the picture cards you have made and ask each child individually to name each item. Encourage the children to listen carefully to the first sound on the tape and explain that you are going to ask them what they think it is. Play the first sound. You might need to repeat this. Ask the children if they can work out what makes that sound. If the children are having difficulty in identifying a sound, display the picture cards to help them guess.

Tell the children that you are going to set up the tape recorder for the children to use themselves. Ask one child to operate the tape recorder and invite the other children to guess – with the help of the clues on the cards – what item has made the recorded sound.

Support
Limit the number of items used for younger children. Let them listen to the items first. Use the picture cards to support them.

Extension
Invite older children to listen to the recorded sounds without looking at the picture cards. Are they abe to identify each item and describe what it does.

CHAPTER 3

Mathematical development

Enhance children's mathematical development with the activities in this chapter. They include controlling a programmable toy to knock down skittles, recording patterns in the environment using a digital camera and weighing different items using electronic scales.

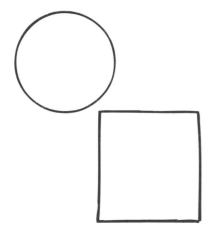

SHAPES AND SIZES

Learning objective
To use language such as 'circle' or 'bigger' to describe the size of shapes.

What you need
Computer system and art package such as *Splosh* (Kudlian), *Primary Colours* (Granada Learning) or *Painter* (Black Cat); colour printer; A4 paper; sets of 2D shapes; the photocopiable sheet on page 67.

Preparation
In advance, prepare practical activities for sorting, recognizing and naming shapes. Copy the photocopiable sheet for each child. Set up the computer system and printer and load the art package.

What to do
Gather the children around the computer. Show them the 2D shapes and invite them to name them. Hold up a shape and ask individual children to find a shape that matches. Ask, 'Can you find a shape that is different?', 'Which shape has three sides?', 'Is there a shape that has two long sides and two short sides?'. Invite the children to draw the shapes in the air and on the floor. Encourage them to discuss the attributes of each shape.

Show the children how to use the art package. Point out the different features, for example, the size of brush and how to choose different colours. Draw different shapes and ask the children to identify them. Then show them how to use the 'fill' option to use different colours to fill the drawn shapes.

Encourage the children to make different shapes and to use a range of colours. Organize them into pairs to work on the computer. Sit with each pair and help them to use the mouse and program. Ask them to identify the shapes and use mathematical language such as 'larger', 'smaller' and 'the same as'. The children can then save and print out their work. Use a copy of the photocopiable sheet on page 67 to show what each child has achieved.

Support
Give younger children plenty of opportunities to explore the shapes.

Extension
Use the gridlock feature in *Splosh* (Kudlian) to explore patterns using different shapes and colours.

A COOL PLACE

Learning objective
To use everyday words to describe position.

What you need
Access to a fridge; a shopping bag; a selection of food such as cheese, meat, tin of baked beans, eggs, soup, packets of cereal, rice, tomato ketchup, lettuce, tomatoes, crisps, apples, butter, yoghurts, bread, milk and biscuits.

Preparation
Place a selection of the different food on the shelves in the fridge and place other items in a shopping bag.

What to do
Take the children to look at the fridge. Ask, 'What is a fridge for?', 'Is it warm or cold inside?'. Explain that a fridge has to be plugged in to the electricity and switched on to keep it cold.

Look inside the fridge together. Point out the temperature control, special compartments for the eggs and milk and the salad tray. Invite the children to describe what else they can see inside. Ask, 'Where is the food?', 'How many shelves are there?', 'What is on the first shelf?', 'Are the yoghurts on the middle shelf or the bottom shelf?', 'Where are the eggs?'. Explain that the shelves are used to store food until it is used.

Show the children the shopping bag and display some of the food items. Invite the children to look at the items and decide if it should be kept in the fridge or not. Encourage the children to explain why some items do not need to be stored in the fridge, for example, baked beans that are sealed in tins. Point out other ways of storing food, for example, in the freezer, in cupboards, in tins or plastic containers. Invite individual children to take other items of food from the shopping bag and place in the fridge if that is where they should be stored. Instruct the children to place food 'on the second shelf', 'next to the milk', 'behind the yoghurt', 'in front of the lettuce' and so on.

Support
Choose four or five items and place them in the fridge. Ask younger children to move them to different places, for example, put the tomato 'next to the cheese'.

Extension
Ask older children to remember where things are placed. Change one or two items around and see if they can say which have changed places.

GROUP SIZE
Any size.

TIMING
Ten minutes.

HOME LINKS
Encourage the children to share the rhymes and songs with family and friends. If possible, organize a loan system for the children to take home their favourite tapes of songs and rhymes. (Remember that some children may not have access to a tape recorder.) Ask parents and carers to help their children to find more counting rhymes and songs and add these to the collection.

MULTICULTURAL LINKS
Make a collection of multicultural number rhymes and counting songs.

CAN YOU COUNT?

Learning objective
To say and use number names in order.

What you need
A selection of counting rhymes and songs such as 'One, Two, Three, Four, Five', 'One, Two Buckle My Shoe', 'One Elephant Went Out to Play', 'One Man Went to Mow' and *Counting Rhymes* compiled by John Foster (OUP); tape recorder and tape.

Preparation
Choose a selection of counting songs and rhymes and make a recording of some of the children's favourites.

What to do
Gather the children together on the carpet explain that you are setting up the tape recorder. Show them the tape recorder and point out the case, label, reels and tape. Explain that the tape passes from one reel to the other as it is being played. Insert the tape. Point out the 'play' and 'stop' buttons on the tape recorder. Ask, 'What happens when you press the 'play' button?'. Show the children the volume controls and turn the volume up and down.

Play two or three of the songs or rhymes on the tape recorder. Explain that someone has recorded these to help children to have fun while they are learning to count. Talk about some of the counting rhymes and songs that the children enjoy, for example, 'Five Little Speckled Frogs' or 'Ten Little Monkeys Jumping on the Bed'. Help the children to say or sing one of their favourite rhymes.

Suggest that the children might like to make a collection of their favourite counting rhymes and songs. Choose a rhyme to say and then explain how you are going to tape the children when they repeat it. Invite the children to listen to their recorded rhyme. Ask, 'What should we do to make it sound louder?', 'How can we stop it playing?'.

Tell the children that you are going to set up the tape recorder so that they can record and listen to themselves counting. Remind them how to use the tape recorder.

Support
Help younger children to operating the tape recorder. Encourage them to join in with the rhymes and songs.

Extension
Ask older children to add more rhymes and songs to share with others. Encourage them to record, rewind and play their recordings.

GROUP SIZE
Any size.

TIMING
Ten to 15 minutes.

DISPLAY IT

Learning objective
To recognize numerals 1 to 9.

What you need
A selection of children's calculators; thick coloured card; felt-tipped pens; scissors; the photocopiable sheet on page 69.

Preparation
Cut up the coloured card and make several sets of 'calculator cards' to represent a calculator. Write some numbers in the display space. Use one-, two- and three-digit numerals. Copy the photocopiable sheet on page 69 for each child.

What to do
Gather the children together and sit with them around a table. Place the selection of calculators on the table. Ask the children if they know what they are. Invite the children to choose a calculator and to look at it carefully. Point out the 'off'/'on' switch, the display space and the number keys. Help the children to switch on their calculators. Encourage them to play with the different keys and see what appears in the display space.

Invite the children to press specific keys, for example, numbers 1 to 5. Ask, 'Can you find the numbers 9, 0 and 4?', 'Can you fill the display with 9s?', 'How many 9s are there altogether?'. Ask each child to show you his/her set of 9s. Check to see if everyone has displayed the same number of 9s. Repeat the activity using different numbers. Make a sequence of two numbers and ask the children to carry on the pattern. Invite them to press the keys in order from 1 to 5. Ask, 'What can you see?'.

Introduce the calculator cards and explain that you would like the children to read the number on the card and copy it into the calculator. Check that the children can recognize the numerals and can find them on the calculator. Give each child a copy of the photocopiable sheet on page 69 and ask them to add some of the features.

Support
Give younger children plenty of time and support to explore the calculator and find out what happens when they press the keys. Focus their attention on the numbers 1 to 5.

Extension
Encourage older children to make their own number sequences. Suggest that they tell their friends which numbers to press, then see if they have the same numbers displayed on the calculator.

HOME LINKS
Ask the children to look for other items that show numbers at home, for example, a clock-face, video-recorder, television remote-control, computer keyboard and bathroom scales.

PROGRAM IT

Learning objective

To use mathematical ideas to solve practical problems.

What you need

Six empty, plastic two-litre fizzy drinks bottles; brightly-coloured wrapping paper; coloured sticky tape; sand; programmable toys such as Pip or Roamer; paper; clipboard; pencil; large space.

Preparation

Decorate the plastic bottles with wrapping paper and sticky tape to make skittles. Fill with a little sand to make them more stable. Make sure that the programmable toy is fully charged or has new batteries. Set out the skittles in the large space in a three, two, one formation. Use sticky tape to make a starting line. This should be a short distance from the 'skittles'.

What to do

Gather the children together near the activity area and ask them if they have visited a bowling alley and seen people knocking down skittles. Show the children the programmable toy and remind them how to control it to move forwards or backwards a number of steps.

Explain to the children that they are going to help you program the toy to knock down the skittles. Ask, 'How many steps do you think it will take to make the toy reach the skittles?'. Program the toy, and watch it move, then ask, 'Has it reached the skittles?', 'How many more steps does it need to move?', 'Has it gone too far?'. Invite the children to guess how many skittles you can knock over in one go.

After several demonstrations, invite individual children to take turns to try to knock down the skittles. Support them in estimating how many steps it takes to reach the skittles. Encourage the children to record the number of skittles knocked down each turn on a piece of paper on a clipboard.

Support

Limit the number of skittles to three for younger children and program the toy to reach the skittles.

Extension

Organize older children to work in pairs. Encourage them to estimate, check and refine their programs using mathematical language such as 'more than', 'less than', 'too far', 'too short'.

GROUP SIZE
Any size.

TIMING
Up to ten minutes.

HOME LINKS
Ask the children to find patterns in and around the home and, if possible, to bring examples into the setting to make a 'pattern' display.

MULTICULTURAL LINKS
Look at patterns used in different cultures. Support children to represent these patterns using different media.

PATTERNS, PATTERNS

Learning objective
To talk about, recognize and recreate simple patterns.

What you need
Digital camera; computer system; *Powerpoint*; a selection of everyday items such as wrapping paper, wallpaper, curtain fabric, floor tiles, wall tiles and clothes; string; construction toys; coloured card; felt-tipped pens; Lego, paints; play dough.

Preparation
Take pictures with a digital camera of patterns found in everyday life, for example, on wrapping paper, wallpaper, curtain fabric, brick walls, floor and wall tiles, fences and garden displays. Set up the computer system. Load *Powerpoint* and select the 'Pattern' slide show.

What to do
Gather the children around the computer. Explain that you have been collecting pictures of patterns. Show the children the slides. Ask them to look carefully for the pattern in each slide. Ask, 'Is it a straight pattern?', 'How many times is the pattern repeated?'. Encourage the children to think where they might have seen the pattern, perhaps inside or outside. Invite the children to recreate some of the patterns using different media, for example, string, construction toys, coloured card, felt-tipped pens, Lego, paints and play dough.

Take the children on a 'pattern search' around the setting. Use the digital camera to record the children's discoveries such as brick walls, slabs and fencing. Add these pictures to the slide show. Demonstrate to the children how to operate the slide show using the mouse and encourage them to explore the slides. Observe the children to see if they can point out the patterns and tell each other where these can be found. Ask the children to carry out a 'pattern watch'. Invite them to look for different patterns on their way home and in the neighbourhood. If possible, take pictures of some of their 'finds' and add to the slide show.

Support
Younger children may need more activities in which they make their own simple patterns before they are able to recognize patterns in the environment.

Extension
Encourage older children to work in pairs and to create their own patterns using real objects.

HOW FAR?

Learning objective
To begin to use the vocabulary involved in adding and subtracting.

What you need
Programmable toy (Pixie, Pip or Roamer); floor space; extra adult support; two empty, plastic, two-litre fizzy drinks bottles; sand or pebbles; brightly-coloured plastic tape; string or strong thread; pale fabric; felt-tipped pens; clipboard; paper.

Preparation
Fill the containers with sand or pebbles and decorate with tape. Clear a floor space. Make a banner saying 'Finish'. Use tape to make a starting line. Place the bottles, adjacent to each other at a suitable distance opposite the starting line. Attach the banner between the bottles. This will vary according to the programmable toy that you are using. Make sure that the number of 'steps' from the start to finish – the bottles – is appropriate for the children's counting ability. Ensure sure the programmable toy is fully charged!

What to do
Gather the children together near the 'racetrack' and ask them to sit near the starting line. Explain that they need to make the programmable toy move from the starting line and to go under the 'Finish' banner.

Demonstrate how to operate the programmable toy. Make sure that the children understand how to program the toy to move forwards and backwards a number of steps. Next, invite the children to take turns to operate the toy and to count the number of steps taken. Ask, 'How many steps have we taken?', 'How many more do we need to make to reach the finish?'. Write the number of steps down on the clipboard. Ask, 'How many did it take altogether?'.

Support
Limit the distance for younger children and help them to count the number of steps taken.

Extension
Encourage older children to record their own number of steps and to compare the steps taken on each child's turn. Ask, 'Were there more or fewer steps on your turn than on Luke's turn?', 'What happens if we change the distance between start and finish?'. Make the distance longer and encourage the children to estimate the number of steps it will take. Ask, 'Was the number of steps greater than or less than your estimate?'.

WEIGH IN

Learning objective
To recognize numerals 1 to 9.

What you need
Electronic scales, a selection of items such as fir cones, shells, buttons, pegs, cubes, plastic straws; adult support; the photocopiable sheet on page 70.

Preparation
Place the items together on a table. Copy the photocopiable sheet on page 70 for each child.

What to do
Show the children the electronic scales. Invite them to suggest what they are and how they are used. Demonstrate how to turn the scales on and off. Show the children the different parts of the scales, for example, the display, 'on'/'off' switches and where items are placed for weighing.

Choose some items to weigh one at a time. Point to the digital display and make sure that the children notice how the numbers change as items are placed on the scales. Can the children recognize some of the numbers? Ask, 'What happens to the display when some items are taken off?'. Invite each child to weigh out some of the items. Ask, 'Can you use the 'on'/ 'off' switch?', 'Can you recognize some of the numbers on the display?'.

Allow the children time to weigh out different items on the scales and ask them to observe what happens to the display. Encourage them to find other items in the room to weigh themselves. Ask, 'Which is the heaviest item?', 'Which is the lightest?'. Invite the children to estimate by holding items in their hands then check to see if they are correct by weighing them on the scales.

Give each child a copy of the photocopiable sheet on page 70 and ask them to record the item they weighed and what was shown on the display.

Support
Help younger children by asking them to look at the numbers on the digital display and then to find the same numbers on a number line.

Extension
Ask older children to record the numbers showing on the display. Can they make the numbers using straws?

CHAPTER 4

Knowledge and understanding of the world

The following activities concentrate on the area of Knowledge and understanding of the world and help children to appreciate information and communication technology in their own surroundings. Ideas include role-playing a fast food shop and taking photographs of buildings in the local environment.

GROUP SIZE
Four to six children.

TIMING
Ten to 15 minutes.

HOME LINKS
Invite the children to go on an ICT hunt at home and suggest that they make some observational drawings of the things that they find.

MULTICULTURAL LINKS
Show the children some examples of technology used by other people in their working lives such as fax machines, movie cameras, microphones, information displays and conveyor belts.

ICT HERE, ICT THERE

Learning objective
To find out about and identify the uses of everyday technology.

What you need
Access to technology items in the setting such as a photocopier, telephone, alarm system, cooker, microwave, kettle, washing machine, tape recorder, computer, printer, television, video-recorder and radio; the photocopiable sheet on page 71.

Preparation
Before carrying out the activity, look around your setting to see what technology is available. Enlarge the photocopiable sheet on page 71 to A3 size. Tell other staff that you are going to take the children on an ICT hunt around the setting.

What to do
Display the A3 photocopiable sheet on a wall or easel. Ask the children to come and sit with you on the carpet. Show them the photocopiable sheet and ask them to identify some of the items. Encourage individual children to describe how each of the items are used. Say that we use lots of items now to help us do things that were not available many years ago, for example, before washing machines were invented people had to wash clothes by hand and people did not have televisions or video-recorders.

Tell the children that you are going on a special hunt to find items around the setting that are seen and used everyday. Go over the items on the photocopiable sheet and encourage the children to look out for these as you walk around.

Take the children on the ICT hunt, pointing out different items. Ask, 'What is it?', 'How is it used?'. Return to the carpet area and mark off the things on the photocopiable sheet that were found on the hunt. Encourage the children to think where they saw the items and how they are used.

Support
Limit the number of items in the search. Cut out pictures of items from magazines and ask younger children to match these on the ICT hunt.

Extension
Give pairs of older children a copy of the photocopiable sheet and a clipboard. Ask them to mark items they find on the ICT hunt themselves.

WHAT, WHY, HOW?

Learning objective
To find out about and identify features of objects.

What you need
Access to a cooker; scissors; glue; spreaders; magazines; catalogues; felt-tipped pens; paper; pencils; the photocopiable sheet on page 72.

Preparation
Collect and cut out pictures of different appliances, including cookers. Copy the photocopiable sheet on page 72 for each child.

What to do
Gather the children near the cooker. Encourage them to say what it is and how it is used. Explain that a cooker has different parts that can be used for cooking, for example, the grill, the hot plates and the oven. Point out these features to the children. Show them the temperature control, timer, clock, shelves and other features. Encourage the children to think about things that might be cooked in the oven and things that are cooked on hot plates. Tell the children about health and safety issues such as not touching the cooker rings, opening doors or lifting and moving pans.

Invite the children to look at the pictures of appliances and find a cooker. Ask them to see if it has the same features as the cooker that they have seen. Give each child a copy of the photocopiable sheet. Ask them to choose a picture of a cooker and to stick it onto the sheet. Ask the children to talk about the special features that their cooker has and suggest that they write these down or act as scribe for them.

Support
Help younger children to identify the main parts of the cooker and encourage them to say what they do.

Extension
Ask older children to compare two different cookers to see if they have the same features.

ICT EVERYWHERE

Learning objective
To find out about and identify features of objects.

What you need
Role-play area set up as a fast food shop; play till; till rolls; old receipts; thick cardboard; felt-tipped pens; baseball caps; aprons or overalls; memo pads; pencils; cartons; chairs; tables.

Preparation
Set up the role-play area as a fast food shop. Provide a long table for the 'counter'. Make 'Special offer' and 'Menu' signs. If possible, before carrying out the activity, take the children on a visit to a local fast food shop.

What to do
Gather the children on the carpet. Remind them about their visit to the fast food shop. Ask 'What did you see?', 'How did people choose their food?', 'How did the assistant take the food orders?', 'Did you see any people cooking the food?', 'Could you buy hot and cold food?', 'How did the shop keep items cold?', 'What happened when the food was ready?', 'How did people know what they ordered and how much it cost?'. How did people pay for their food?'.

Show the children the till receipts and point out some of the items. Explain that when the assistant takes an order, he uses special keys on the cash register to show the food that has been chosen. This special key is coded to give the price and item. The information goes on the till receipt. It helps to find out what the favourite meal is, what to order for the fast food shop and keeps a record of what is sold each day.

Encourage the children to tell you about diferent fast food shops that they might visit or see in the local area, for example, Chinese and Indian takeaways, fish and chip shops, pizza shops, burger and sandwich bars. Organize the children into small groups to play in the role-play area.

Support
Observe younger children's play and intervene to help them to understand technology by asking questions and posing requests.

Extension
Ask older children to make menus, special offer posters and their own order books.

COPY IT

Learning objective
To ask questions about why things happen and how things work.

What you need
Access to a photocopier; white A4 paper; pencils; black crayons or felt-tipped pens.

Preparation
Familiarize yourself with the facilities of the photocopier. Cut each sheet of A4 paper into eight pieces.

What to do
Organize the children into pairs. Explain that they are going to do some observational drawings of each other using black crayons or felt-tipped pens. Tell the children that they need to look carefully at their partner's face and notice the shapes of their eyes, nose, mouth and eyebrows. Point out differences in hairstyles and length of hair. Give out the paper and mark-making resources. Allow the children time to observe and draw their portraits. Support the children by encouraging them to refine their drawings and add more details. Take time to look at individual drawings, pointing out special features. Put all the drawings together on one sheet and tell the children that you will use the photocopier so that the children can each have a copy of the drawings.

Take the children to the photocopier. Show them the keypad, where the paper is stored, where the blank paper enters and where their drawings are placed for copying. You might need to set the photocopier on the darkest setting. Do one photocopy and show the children the copy. Ask, 'Are the drawings the same?', 'How many more copies are needed for everybody to have one?'. Reset the copier to give extra copies.

Support
Organize younger children into smaller groups. Encourage them to look closely at each other. Help the children with details by focusing their attention on parts of the face.

Extension
Give older children mirrors and invite them to draw their own portrait. Encourage them to observe individual features closely in the mirror. Help them with the finer details such as the shape of the nose, eyebrows and mouth. Encourage them to describe the texture of their hair and experiment with different techniques to represent it.

GROUP SIZE
Four to six children.

TIMING
Ten to 15 minutes.

OUR PLACE

Learning objective
To observe and find out about the place where we live.

What you need
Extra adult helper; camera; film; photograph album; photo corners; flip chart; old photographs showing familiar places such as local buildings and play areas; indoor and outdoor photographs; the photocopiable sheet on page 60.

Preparation
Discuss the activity with the adult helper. Take a set of photographs of the local area and arrange for these to be developed. Tell others in the setting that you will be taking photographs.

What to do
Gather the children together. Read the poem on the photocopiable sheet and discuss it with the children. Explain that you have taken photographs in the local area. Show them the photographs and encourage them to look carefully and to identify the buildings. Ask, 'Where is this?', 'How do you know?', 'What is inside?', 'Why do we go there?'.

Ask the children to suggest what they would like to take a photograph of in the environment and list these suggestions on the flip chart. Show the children the camera. Tell them that to take photographs the camera has to have a film inside. Point out the viewfinder, lens, and shutter. Explain that the photographs need to be developed before they can be looked at.

Now go around the setting with the children, inside or outside, and take the photographs on the list. Ask, 'Do we need to be far away?', 'Can we get everything in the picture?', 'Which is the best way to hold the camera?'. Vary the photographs by taking different activities and locations, including children and parents arriving and leaving, playtime, any staff and visitors. Repeat the activity with different groups of children until the film is finished. If possible, arrange for two sets of prints to be developed. Show the children the developed photographs and negatives and invite them to suggest captions for a display.

Support
Help younger children to decide on photograph locations. Try and include them in some of the photographs.

Extension
Help older children to take the photograph of their 'chosen' location themselves.

HOME LINKS
Encourage parents and carers to come in to see the photographs of the setting. Suggest that their children explain what is happening in the photograph and where it was taken.

MULTICULTURAL LINKS
Invite the children to bring in photographs of their own special places relevant to their cultures.

GUESS WHAT?

Learning objective
To find out about and identify features of objects.

What you need
Catalogues; magazines; scissors; glue; spreader; thick card; the photocopiable sheet on page 73.

Preparation
Cut out a selection of pictures of technological items, for example, televisions, cameras, camcorders, tape recorders, radios, cookers, fridges, freezers, microwaves and washing machines. Mount each picture onto card and cut each one into three or four pieces. Prepare the copies of the photocopiable sheet on page 73.

What to do
Place the 'technology cards' on a table and the invite children to sit around the table. Show them the copy of the photocopiable sheet on page 73 and point out some of the items. Ask individual children to identify some of the other items and encourage them to describe how they are used.

Show the children the 'technology cards' and explain that the pieces go together to make pictures of things that help us in the home. Invite the children to fit the pieces together to complete the pictures. Suggest that the children look for clues, for example, whether it has a hotplate, monitor or oven. Encourage them to identify special features for the objects. Ask them to describe how they work and what they are used for.

Support
Limit the quantity of technology cards used for the activity. Help younger children to identify the items.

Extension
Help older children to make their own set of technology cards. Give them opportunities to play with their cards, then invite them to share these with their friends.

GROUP SIZE
Whole group for introduction, then groups of up to four children for activity.

TIMING
Ten to 15 minutes.

MEMORIES

Learning objective
To find out about past and present events in their own lives.

What you need
A large wall and table-top for display; Blu-Tack; photographs of the children and yourself at different stages of life and on special occasions such as weddings, holidays, school, as a baby, with family and friends; photographs of people engaged in different activities such as festivals, role-play, sports activities, visits, indoor and outdoor activities, group and individual activities.

Preparation
A few days before you carry out the activity, ask parents and carers to send in pictures of their children, for example, as babies, on special occasions, with family members or on visits.

What to do
Gather the children together on the carpet. Show them your selection of photographs and talk about them. Invite the children to describe what they can see. Ask, 'What is happening?', 'How do you know?', 'Was the photograph taken recently or a long time ago?', 'How can you tell?'.

Explain that we can share favourite times with friends and remember special times if we record them using a camera. Invite the children to share some of their photographs with the group. Can they order their photographs into ages – starting with them as babies? Can they remember what happened on visits or birthdays? Display the children's photographs for everyone to see.

Support
Encourage younger children to describe what is happening in their own photograph. Ask, 'Where was it taken?', 'Was it recently or a long time ago?', 'Can you point out special people in the photograph?'. Encourage the children to recall the event and share it with other children.

HOME LINKS
Take photographs of the children doing different activities in the setting. Place these in a photograph album with annotated comments and invite the children to take the albums home to share with their parents and carers.

Extension
Invite older children to make a 'timeline' using their photographs. Help them to sort their photographs and suggest that they look for 'clues' to see if it is a recent photograph or one taken a long time ago.

GROUP SIZE
Four children.

TIMING
Five to ten minutes.

WHAT HAPPENS NEXT?

Learning objective
To ask questions about why things happen and how things work.

What you need
A programmable toy such as Pixie, Pip or Roamer; a large space.

Preparation
Make sure that the programmable toy is fully charged.

What to do
Sit with the children in a large space and explain that you are going to be the robot commander and the children are going to be robots. Tell them that you are going to give them special commands to move them around the space. Encourage the children to show you what happens when you command them to move forwards. Command the children to 'stop' and encourage them to suggest and demonstrate how you can tell them to move different distances in the space. Ask individual children to move forwards six steps. Repeat, going backwards, then combine the two commands. Observe the children to note if they understand 'forwards' and 'backwards'. Encourage them to work in pairs and give more commands with different steps. Suggest that the children might like to turn instead of just moving forwards and backwards. Encourage the children to work out their own solutions and think about the language that they need to use. Introduce the idea of 'left 90°' and 'right 90°' by demonstrating the turn.

Invite the children to gather around you on the floor. Show them the programmable toy and remind them how to work it. Point out the keypad. Ask, 'What do you think will happen when you press 'Forward'?'. Encourage the children to show you which direction it will move. Ask, 'How far do you think it will travel?'. Invite the children to try some of their suggestions. Ask, 'Can you move the toy forwards five steps?', 'How do you make it go backwards?'. Encourage the children to increase and decrease the number of steps taken. Program the toy to turn 'left 90°' and 'right 90°'. Help them to make a sequence of instructions, predicting what will happen and then check to see if they are correct.

Support
Program the toy with single commands and encourage younger children to observe what happens. Repeat the instructions, helping the children to use the keypad. Build up the instructions until you are making a short sequence for the children to follow.

Extension
Invite older children to record their instructions and show what happens in words or pictorial form.

HOME LINKS
Ask the children to look at home for items that can be programmed, for example, the video-recorder, washing machine and microwave.

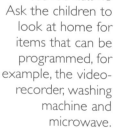

CHAPTER 5

Physical development

Give children the opportunity to develop physical skills by using a range of materials and equipment. Activities include exploring walkie-talkies and baby alarms, playing with remote-controlled cars to understand spacial awareness and following different recipes to make healthy foods.

GROUP SIZE
Two to four children.

TIMING
Up to ten minutes.

HOME LINKS
Ask the children to carry out a 'toy inspection' with any electronic toys that they have at home, telling parents and carers what they have found and reporting back to the group.

TOY INSPECTORS

Learning objectives
To use a range of small equipment; to gain increasing control over an object by touching, pushing, patting and throwing.

What you need
A selection of electronic toys; speaking dolls; soft toys; badges; felt-tipped pens; the photocopiable sheet on page 58.

Preparation
Ask parents and carers if their children would be willing to loan any electronic toys, explaining that they play an important role in children's understanding of technology. Take time to explore what each toy can do. Be aware of health and safety issues especially with small parts and very young children. Make a 'Toy inspector' badge for each child.

What to do
Read the story on the photocopiable sheet to the children and talk about how the toys came to life. Gather the electronic toys together and place in an appropriate area. Divide the children into small groups and explain that there are many electronic toys available and that these toys have some form of control. Hold up one toy and ask the children to describe how to make it perform different actions. Invite individual children to demonstrate different things. Ask, 'What happens if you clap your hands?', 'Does it talk?', 'Where do you press to make it do different things?'.

Organize the children to work in pairs. Suggest that they are 'toy inspectors' and that they need to discover what the toys do. Give all the children a badge and ask them to write their name on it, then to put it on. Encourage the children to think whether each toy would be a suitable toy for a baby, toddler or teenager and to give reasons for their decision. Ask them to sort their toys in order of preference.

Support
Help younger children to explore the electronic toys and to describe what they have discovered. Let them enjoy playing with the toys and encourage them to discover what each one can do.

Extension
Suggest that older children identify each action that the toys make to find out which is the most versatile.

GROUP SIZE

Whole group introduction; two to four children for activity.

TIMING

Five to ten minutes.

ON PATROL

Learning objectives

To handle equipment with increasing control; to collaborate in devising and sharing tasks; to show awareness of space, themselves and others.

What you need

Two sets of children's walkie-talkie receivers; batteries; baby alarm; mobile phone.

Preparation

Set up the baby alarm in the role-play area. Insert new batteries if required.

What to do

Gather the children together on the carpet. Show them the mobile phone and ask if they know why it is used and how. Explain that for many years people have used landline phones but over recent years mobile phones have become much more popular. Show the children the main features and point out that people can keep in contact by phoning and sending text messages. Point out that there has to be a strong signal and that they need to be fully charged, otherwise they might not work.

Explain to the children that there are other ways of keeping in touch. Ask if anyone has a baby at home. Ask if their parents can listen to the baby if they are in another room. Show them the baby alarm and suggest that a couple of children go into the role-play area. Ask them to talk, whisper and sing and see if they can be heard.

Ask the children to think of other people who need to keep in contact with each other as part of their jobs, for example, police, security officers, firefighters. Show them the walkie-talkie sets and demonstrate how they work. Suggest that the children work in pairs and pretend that they are on patrol. They will need to tell each other where they are located, where they are going next and what they can see.

Support

Let an adult or older child work with younger children.

Extension

Encourage older children to give clear information about their location. They can test the range of the walkie-talkies.

HOME LINKS

Ask the children to find out about other forms of communication used at home such as fax machines, telephone answering machines and e-mail.

MULTICULTURAL LINKS

Give support to children with English as an additional language by involving bilingual staff or parents and using the children's mother tongue.

GROUP SIZE
Four children.

TIMING
Ten to 15 minutes.

PATHWAYS

Learning objectives
To show awareness of space; to collaborate in devising and sharing tasks.

What you need
Programmable toys such as Pixie, Pip and Roamer or remote-controlled cars; coloured masking tape or large plastic or wooden bricks; large floor space; two empty one-litre plastic drinks containers; sand; thin cardboard; sticky tape; felt-tipped pens.

Preparation
Make captions saying 'Start' and 'Finish'. Partially fill the two empty plastic bottles with sand. Attach each of the captions to a plastic bottle. Make a 'racetrack' layout on the floor using the coloured tape or large bricks. Set up a 'Start' and 'Finish' area.

What to do
Gather the children near to the racetrack. Explain that they are going to work in pairs to have a race. Show the children the controls and remind them how to operate the programmable toys or remote-controlled cars. Place one of the vehicles on the starting grid and demonstrate how to move it around the racetrack. Support the children in completing a 'test run' by moving their vehicles to do a couple of circuits of the track.

   Next, organize the children into pairs and explain that they are going to take part in the 'Grand Prix'. Start the race. Encourage each pair to complete the circuit with the other children giving support. Allow opportunities and time for each pair to race their vehicles a few times around the racetrack.

Support
Make the layout of the racetrack very simple for younger children so that it is easier to navigate.

Extension
Make the layout of the racetrack more complicated for older children or let them make up their own. Add a timing element by using sand-timers.

HOME LINKS
Give each child a copy of the photocopiable sheet on page 74. Place each sheet in a plastic wallet with a dice and four different-coloured counters. Explain how to play the game to parents and carers and suggest that the children have a race with them at home. Encourage the children to play other dice games such as *Snakes and Ladders* or *Ludo* at home with parents and carers.

CHEF'S SPECIAL

Learning objectives

To recognize the importance of keeping healthy and those things which contribute to this; to recognize and promote understanding that foods from a range of cultures can be enjoyable and form a healthy diet.

What you need

Cooker; electronic scales; a selection of recipes from different cultures; cooking ingredients; utensils; preparation facilities; aprons; adult helpers.

Preparation

Invite parents and carers to provide a selection of simple recipes that can be shared with the children over several weeks. Select a different recipe for each group, for example, samosas, crêpes, fruit salad, vegetable soup, pizza or noodles. Gather the ingredients together. Invite a parent or carer to come in and work with the children. (NB Be aware of any food allergies or special dietary requirements.)

What to do

Explain to the children that people do not all like or eat the same foods. Ask individual children to share their own favourite food with the group. Explain that there are many different foods that we can eat and enjoy and that you are going to organize the children into groups to try out some new and different recipes.

Gather a group of children together and introduce the recipe. Give some information about the country of origin, for example, pizza from Italy, crêpes from France or noodles from China. Show the children the ingredients to be used and some of the cooking utensils you will need.

Make sure that all the children are aware of personal hygiene, for example, washing hands, wearing aprons and not eating uncooked food. During the preparation, discuss foods, cooking methods and encourage the children to think about healthy eating. Encourage the children to help weigh out ingredients using electronic scales and to be aware of the use of the cooker for different parts of the food preparation.

When the food is ready, assemble the group together and discuss the result. Ask, 'What does it look like now it is cooked?', 'When would it be eaten – at breakfast, lunch, tea or dinner?', 'Would it be suitable for a picnic?'. Organize a tasting session and discuss the different flavours. Ask, 'Is it sweet or savoury?', 'Is it spicy?'.

Support

Work on a very simple recipe that takes only a short time to prepare and make in order to hold younger children's interest.

Extension

Encourage older children to suggest appropriate food to try and to bring in any recipes that they have used at home.

THE CAR WASH

Learning objectives

To show awareness of ourselves and others; to operate equipment by means of pushing and pulling movements; to collaborate in sharing tasks.

What you need

Outdoor area; play vehicles such as cars, trikes, bikes, scooters; plastic buckets; sponges; washing-up liquid; water; chamois leathers; different-sized brushes; protective clothing; thick cardboard; short bamboo canes; plastic tubs; sand; felt-tipped pens; chalk; traffic cones; playground chalk or washable paint; large paintbrushes; adult support.

Preparation

If possible, take the children on a visit to a car wash to see what happens; if not take photographs or ask the children to look out for car washes when they are out with parents. Assemble the role-play items. Make signs saying 'Park here', 'Car wash', 'Welcome', 'Entrance' and 'Exit' on pieces of cardboard attached to bamboo canes. Put some sand into small buckets and stick the canes into the sand. Set up the 'car wash' in a suitable place outside. Mark out or paint parking spaces for the outdoor play vehicles. Number these from 1 to 6 with extra unnumbered spaces.

What to do

Gather the children together in a space. Ask the children to share their observations and experiences at the car wash or discuss the photographs. Outline what happens at the car wash. Explain that car washes are programmed to wash vehicles in different ways and that customers have choices about the type of wash they require. Invite the children to think about the equipment used (sprayers and revolving brushes) and the sequence of choosing the wash, paying, driving in, having the car washed, shampooed, rinsed and then driving out of the car wash.

Suggest that the children have their own car wash for their outdoor play vehicles. Take the children outside in small groups and point out the layout. Read the signs and ask for suggestions where to position them. Give the children rules about wearing protective clothing, using water and washing equipment and cleaning the vehicles. Invite the children to sort themselves into customers and car washers. Allow them the opportunity for play and only intervene when appropriate.

Support

Help younger children in their role-play, asking questions or inviting them to perform certain roles.

Extension

Encourage older children to make further resources for the car wash such as information signs, tokens and waiting area.

GROUP SIZE
Any size for introduction; pairs for activity.

TIMING
Up to ten minutes.

HOME LINKS
Give each child a copy of the photocopiable sheet on page 76 to take home. Encourage the children to use their pencils to find a route around the maze for their mouse to get to the mouse hole. Show them how to hold the pencil carefully and correctly to help with pencil control.

MOUSE TRAILS

Learning objectives
To handle tools with increasing control; to develop hand–eye co-ordination.

What you need
Unattached computer mouse; mouse mat; computer system with mouse attached.

Preparation
Set up the computer system. Place the mouse and mouse mat nearby.

What to do
Gather the children together near the computer. Hold up the unattached mouse. Point out the special features such as the buttons, ball and cable to the computer. Invite the children to describe and demonstrate how to use the mouse. Explain that when you move the mouse, the ball underneath moves and sends special messages to the computer. The buttons are used to make choices when using the computer. Demonstrate how to move the mouse on the special mat.

Invite the children to watch the computer monitor and notice what happens when you move the mouse. Ask, 'What did you see?', 'What happens on screen when I move the mouse up and down?', 'How do I make it go to the left then right?'.

Let individual children move the mouse. Ask the children to follow the mouse trail around the screen. Can they make the mouse move in a circle or square? What do the children think will happen when you move to the edge of the screen?

Organize the children to work in pairs and encourage them to take turns moving the mouse around the screen.

Support
Help younger children to manoeuvre the mouse around the screen by holding your hand over the child's hand. Then use your finger to point to the screen. Gradually move your finger around the screen challenging the child to follow it with the pointer by moving the mouse.

Extension
Use an art package or word-processing package and ask older children to choose different options using the mouse buttons.

SWITCH AND GO

Learning objectives
To use a range of small and large equipment; to understand that equipment has to be used safely.

What you need
A selection of torches; tape recorders; tapes; CD players; CDs; radio; camera; alarm clock; electronic scales; microwave oven; adult support.

Preparation
Set out the items together on a table. Give clear information to the adult about the objectives of the activity.

What to do
Gather the children together and point out the items on display. Encourage them to identify the items one by one and to describe how they are used in everyday life. Invite the children to look closely at one of the items, for example, the microwave oven. Ask, 'Where are the controls?', 'How do they work?', 'Does it use batteries, gas or electricity?'. Encourage individual children to explain how they have seen a microwave being used. Explain that it is important to use it safely and correctly and with adult help. Demonstrate how to use the microwave oven and point out the different features.

Give the children the opportunity to explore the other items. Observe how they use them. Ask, 'Can you switch them on and off?', 'Can you help each other?', 'Can you show someone else how it works?. Invite individual children to demonstrate and describe to the group how to operate each item.

Support
Work with individuals or pairs. Limit the quantity of items and help younger children to operate each one.

Extension
Encourage older children to work in pairs. Ask one child to give instructions to another to operate each item then to change over.

GROUP SIZE
Any size.

TIMING
Up to 20 minutes.

MOVEMENT AND MUSIC

Learning objectives
To move with control and co-ordination; to move with confidence, imagination and safety.

What you need
A tape recorder; music tape; large open space.

HOME LINKS
Make tapes of music, songs and rhymes with actions that can be shared with parents and carers.

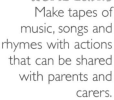

MULTICULTURAL LINKS
Be sensitive to the requirements of modesty in some cultures and religions with regard to PE clothing. Include a range of music from different cultures.

Preparation
Select a piece of music for the activity. Ask the children to bring in loose and comfortable clothing, for example, shorts and T-shirts.

What to do
Set up the tape recorder in the hall or large space. Gather the children together and support them in changing into their PE clothes, with either pumps or bare feet. Encourage the children to manipulate fastenings and to help each other to change. Point out the need for loose clothing to move freely and to be able to use all the space available.

Invite the children to listen to a section of the music. Ask, 'What does it sound like?', 'Is it fast or slow?'. Explain that you are going to play it again and you would like them to move around the room in any way they want to. Stop the music. Ask individual children to show their movements to the group. Next, invite the children to move in different ways, for example, backwards, sideways, walking, jumping, running, hopping. Encourage them to listen to the music again and change the speed and direction, making sure that they avoid obstacles and each other. Encourage the children to explore different ways of moving. Ask individual children and groups to demonstrate their actions.

Support
Demonstrate new skills and actions to younger children. Give them the opportunity to listen to the music and to move freely in the space.

Extension
Encourage older children to combine movements and gestures to make sequences. Invite them to work in pairs and small groups.

CHAPTER 6

Creative development

Focus on Creative development with the following activities. Children can explore colour using an art package, express and communicate ideas through role-playing a garage, utilizing resources to make a stage area and perform their favourite pop group and role-play using technology at a supermarket.

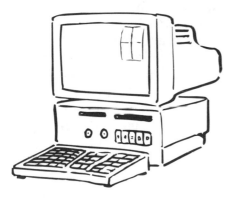

HOME LINKS
Invite the children to create pictures, by drawing, painting or collage, or using computers at home. Give each child a copy of the photocopiable sheet on page 77 to display their work.

MULTICULTURAL LINKS
Collect examples of art from different cultures to add to the gallery.

ART GALLERY

Learning objectives
To explore colour, shape and form in two dimensions; to use ideas involving fitting, overlapping, in, out and enclose; to choose particular colours to use for a purpose.

What you need
Computer system; colour printer; art package such as *Splosh* (Kudlian), *Primary Colours* (Granada Learning) or *Painter* (Black Cat); coloured A4 paper; thick coloured card; a selection of paints; an area for wall display.

Preparation
Place the coloured paper near the printer. Cut out different-sized frames from the thick coloured card. Give the children plenty of experience using 'real' paints, mixing colours and exploring different marks and movements on paper. In advance, introduce the children to the facilities of the art package. Show them how to use the different tools for drawing, choose different palettes or colours, clear the screen, fill the screen, delete or save their work and how to print it out (with support).

What to do
Organize the children into small groups near the computer. Remind them how to use some of the facilities of the program. Invite individual children to demonstrate some of these, for example, how to change colours, use drawing tools, delete and clear the screen. Ask the children to try different things and say that you would like them to save their paintings as you want to keep their artwork.

Invite the children to choose and print out one of their 'creations' for an art exhibition. Support the children in saving and printing out their work. Mount and frame their work and display their chosen pictures in the 'art gallery'.

Support
Encourage younger children to explore different ways of making marks, for example, using different sizes of brushes, special effects features and different colours. Help them to save and print out their work.

Extension
Encourage older children to experiment with different techniques. Give them the opportunity to save and print out their own work.

AT THE GARAGE

Learning objective

To express and communicate ideas through imaginative play and role-play.

What you need

A suitable outdoor space and a selection of outside play equipment such as bicycles, cars, trucks and traffic cones; a role-play area; play till; telephone; play cooker and microwave; empty food packets; old newspapers; artificial 'flowers'; desk; pens; paper; till receipts; large cardboard boxes; paint; large paintbrushes.

Preparation

Assemble the resources. Paint the large cardboard boxes to resemble petrol pumps. Set up the role-play area adjacent to the outdoor play area. Set out the play vehicles and traffic cones. Set up the role-play area as a garage shop.

What to do

Gather the children around and talk about the garage. Ask, 'Why do people go to the garage?', 'How do you put petrol in your car?', 'What else might you see at a garage?', 'What happens in the shop?', 'What can you buy?'. Explain the technology that can be found at a garage, for example, the pumps, car wash, cash register, credit card facilities, freezers, fridges and microwaves. Ask, 'How does the cashier know how much to charge for petrol?', 'What happens in the car wash?', 'How do you heat up a pie in the shop?'.

Point out the outdoor and indoor role-play areas to the children. Invite them to pretend that it is the local garage. Ask them to set out the resources themselves. Encourage them to think of other facilities that might be found at the garage. Take time to observe the children at play. Intervene if necessary, but try to remain in the background.

Support

Initially, take the lead with the role-play to encourage younger children to participate. Stay beside the children but gradually take a less active role as they become more confident. Take the opportunity to observe the children's play.

Extension

Encourage older children to show their understanding of technology in everyday life. Invite them to describe what, why and how things happen.

POP STARS

Learning objectives

To use one object to represent another; to use resources to create props to support role-play.

What you need

CD player and CDs (or tape recorder and tapes); large space; selection of recycled materials to represent instruments and microphones; large construction bricks.

Preparation

In advance, find out who are the children's favourite pop groups. If some children do not know any, encourage those with older siblings to volunteer suggestions. Assemble a selection of CDs or tapes by these artists. Collect a variety of recycled materials such as boxes and kitchen-roll tubes. Place the large construction materials in the 'stage' area.

What to do

Set up the CD player. Gather the children together on the carpet and invite them to listen to a piece of music. Show them how to insert and start the disk, and how to use the volume control. Encourage them to identify the singer or group. Ask, 'What instruments do they play?', 'How do they perform on stage?', 'Do they have special dance routines?'. Invite two or three children to perform to the music. Point out that when groups sing 'live' they have microphones to make the sound louder so that the audience can hear. Explain that sometimes microphones are on stands, while others are portable. Also, if people are dancing and moving around there are sets that can be worn on the head with headphones and microphones in front of the mouth. Some children may have seen singers or groups perform on television and videos and others may even have been to live concerts.

Suggest that the children make a stage area and perform as their favourite group. Invite them to use the materials for instruments such as small cereal boxes mounted on a section of kitchen-roll tube to make a microphone and suggest that they plan a singing and dancing routine. Remind them how to operate the CD player and suggest that they choose one or two of the songs. Allow opportunities for the children to carry out their role-play without constant intervention. Encourage some groups to share their performances with other children acting as the audience.

Support

Use mixed age and ability groups. Provide resources for younger children to use such as dressing-up clothes, microphones with stands, headphones and musical instruments.

Extension

Encourage older children to plan their acts and include sequences in their dance routines.

GROUP SIZE
Six children.

TIMING
Ten to 15 minutes.

ROBOT GAMES

Learning objective
To explore shape, form and space in three dimensions.

What you need
Thick card; felt-tipped pens; programmable toys such as Pixie, Pip or Roamer; large foam dice numbered 1–6; large space; coloured bands or badges for each child; adult support.

Preparation
Make six sets of cards with the following instructions: 'forwards', 'backwards', 'right 90°', 'left 90°', 'stop', 'wait a turn', 'miss two turns', 'take two turns'. Ensure that the programmable toys are fully charged or have new batteries. Mark out a track of a suitable length. Make a starting and finishing point at opposite ends.

What to do
Assemble the items for the game. Gather the group together in a space. Invite the children to compete in the 'Robot games' and organize the children into two teams. Provide coloured bands or badges to identify each team.

Remind the children how to program the toys. Explain that the object of the game is to control the robots and see which one is first to go up and down the track. Show the children the cards and explain that each card will give instructions about what to do. Read out the other cards telling the children what they need to do. If the children turn over a 'forwards' or 'backwards' card they need to throw the dice to see how many steps they can take.

Take the cards and place each one face down on the floor. Invite the children to pick a card, read what it says and then control the robot as directed. When the children are confident about how to play the game, the challenge can begin and they can play against each other.

Support
Give younger children plenty of opportunities to move their robots around. Help them to read the cards. Use a dice with spots and help them to count the spots and show them what to do when necessary. Assist the children in programming their robots to follow the directions.

Extension
Encourage older children to make extra directional cards that they can add to the game.

HOME LINKS
Suggest that the children play traditional games such as *Snakes and Ladders* and *Ludo* to develop skills in turn-taking and moving forwards and backwards.

MULTICULTURAL LINKS
Find out about games from different cultures and include these in activities.

SOUND PICTURES

Learning objectives

To use imagination in art and design; to extend children's experience and expand their imagination through music; to compose a sound picture.

What you need

A selection of music on CDs or tapes; CD player or tape recorder; selection of art materials such as paints, pastels, wax crayons, collage materials, paintbrushes, glue, spreaders; different-sized paper; aprons.

Preparation

Select a suitable piece of music. Set up the CD player or tape recorder. Prepare the working area and set out the art materials.

What to do

Gather the group together and invite them to sit in a space on the floor. Explain that you have chosen a special piece of music and you are going to listen to it together. Play the music and ask the children to describe what they have heard and what they think it is about. If, for example, the music is about the sea, ask them why it reminds them of the sea. Encourage them to close their eyes and think about the sea. Ask, 'What does it look like?', 'What happens to the waves?'. 'Has anyone ever seen the sea crashing onto rocks?'.

Play the music again and encourage the children to listen carefully. Ask, 'Can you hear the waves crashing?', 'Does the music get louder and softer?', 'Is it fast or slow?'.

Point out the selection of art materials and invite the children to make a 'sound picture' about the sea. Tell them that they can listen to the music again as they work. Give the children the opportunity to refine their work before it is displayed for all to see. Invite them to suggest captions about their pictures for you to scribe.

HOME LINKS
Invite the children to bring pictures and photographs of the sea from books and magazines or from family holidays.

At the end of the session, gather the children together again and talk about the pictures that they have painted. Invite each child to tell the group what their painting represents.

Support

Collect information books and pictures about the seaside. Use these as a stimulus to encourage younger children to describe what they might see if they went to the seaside.

Extension

Give older children the time and opportunity to keep refining their work and encourage them to add finer details using different resources such as felt-tipped pens or collage materials.

GROUP SIZE
Any size.

TIMING
Ten to 15 minutes.

HOME LINKS
Make copies of the children's 'top ten' songs. Allow the children to take home a tape recorder, if necessary, to share the songs with their parents and carers.

MULTICULTURAL LINKS
Include a range of multicultural songs and rhymes. Ask parents and carers to help you with this and invite them to record some of these for you to share with the children.

OUR TOP TEN

Learning objectives
To begin to build a repertoire of songs; to explore the different sounds of instruments.

What you need
Tape recorder; blank tapes; favourite tape of children's songs; musical instruments such as triangles, drums, chime bars, cymbals, Indian bells, maracas, xylophone, glockenspiel and keyboards.

Preparation
Play the songs tape to the children on different occasions. They will need to know a variety of songs, including nursery rhymes and action songs beforehand. Set up the instruments and tape recorder in a suitable place.

What to do
Gather the children around you and invite them to sing a few familiar songs and nursery rhymes. Explain that you have recorded a selection of songs for them to enjoy. Show the children how to load the tape and operate the tape recorder. Play a couple of songs from the tape and ask them to listen first, inviting them to sing along afterwards.

Ask the children to tell you some of their favourite songs. Ask if they might like to make a recording of their own favourite songs for everybody to share. Suggest a song for them to sing and explain that you will record it onto a blank tape. Play back the song for them to hear. Encourage them to add instruments. Choose individual children to play some of the instruments as the others sing their song again. Record the song once more, reminding the children how to operate the tape recorder.

Organize the children into small groups and let them record their songs, adding music. Encourage the children to sing different songs to make a collection of their 'top ten' songs.

Support
Help younger children to record their songs and nursery rhymes.

Extension
Encourage older children to use a range of instruments and be responsible for recording their own music.

POST OFFICE PLAY

Learning objectives

To use imagination in role-play and stories; to work creatively on a small scale.

What you need

Computer system with CD-ROM drive; the book *The Jolly Postman* by Janet and Allan Ahlberg (Methuen); *The Jolly Postman's Party* and/or *The Jolly Post Office* CD-ROMs (Dorling Kindersley); stationery such as stamps, cards, writing paper and envelopes; adult support.

Preparation

Set up the computer system and familiarize yourself with the CD-ROMs. Read *The Jolly Postman* to the children. If possible, take them to visit a post office or invite a postperson to visit the setting before you carry out the activity.

What to do

Gather the children together, hold up *The Jolly Postman* book and remind the children of the story. Ask the children to describe things they might see in a post office, for example, stamps, electronic scales, postpeople, customers and posters.

Show the children a used envelope and point out the features – name, address, postcode, stamp and marks where the stamp has been 'franked'. Ask the children to describe what happens to a letter when it has been posted. Explain that post has to be sorted by hand or by special machines at a sorting office, and then franked before it is delivered. Show the children the stationery and invite them to compose a short letter or 'write' a card to a friend.

Introduce the *The Jolly Postman's Party* or *The Jolly Post Office* CD-ROM and load it onto the computer. Demonstrate how to use some of the facilities. Suggest that the children might like to work on the program and help in the post office shown on the screen.

Support

Work with pairs of younger children and help them to use the program. Encourage the children to use the mouse and describe what they are doing on screen.

Extension

Invite the children to write their own letters and cards, either using a word-processing or publishing package or by hand.

HOME LINKS

Ask parents and carers to let their children sort the post that arrives at home and write letters and cards to family and friends. Give each child a copy of the photocopiable sheet on page 79 and ask them to match the correct shape of the parcels.

MULTICULTURAL LINKS

Collect stamps from different countries. Invite parents and carers to come in to help the children to write in their own language.

OPEN ALL HOURS

Learning objective
To use imagination in imaginative play and role-play.

What you need
Role-play area set up as a supermarket or shop; cash registers; till rolls; old till receipts; tables; trolleys; shopping baskets; paper carrier bags; play food; electronic scales; empty cardboard containers; display space; large sheets of paper; thick card; felt-tipped pens; overalls or shirts; the photocopiable sheet on page 80.

Preparation
In advance, take the children on a visit to the local supermarket or shop. Make a copy of the photocopiable sheet on page 80 for each child. Set up the role-play area, inviting the children to help in the layout of the shop and displaying the items.

Make labels such as 'Next customer' and 'Today's best buy'. Invite the children to make posters and shop signs. Set up a checkout area using the tables with a 'bar0code scanner' near the till.

What to do
Gather the children together and look at the role-play shop. Ask them to describe what it is like shopping in a supermarket. Ask, 'What type of things do people buy at a supermarket?', 'How do people weigh their vegetables?', 'What happens at the checkout?', 'How do you know how many items you have bought?', 'How do people pay for their shopping?'. Emphasize the role of technology in shopping. Ask the children if they know the differences between a supermarket and a shop.

Organize the children into groups to play, one group at a time, in the supermarket or shop.

Give each child a copy of the photocopiable sheet and invite them to draw items that they might find in the supermarket in the shopping trolley.

Support
Encourage younger children to play together and help them if necessary. Take time to observe the children in their play and assess their knowledge, understanding and creative skills.

Extension
Invite older children to make their own resources to add to the role-play area. Encourage them to weigh out vegetables and fruit.

Pass the message, please

When the telephone rang, Jodie's mum was in the garden, hanging out the washing. Jodie knew that she should call Mum, but she really wanted to answer it herself. She loved telephones, and she knew exactly what to say. She had heard Mum doing it lots of times. So she picked up the phone.

'Good morning,' said Jodie, very properly. 'Can I help you?'

'Good morning,' said a lady's voice. 'I'm calling to let you know that the car will be ready at 2 o'clock. Could you please collect it then?'

Jodie was thinking hard about what to say next, so she didn't listen very carefully to the message.

'That should be fine,' she told the lady, just like she'd heard Mum do so often.

'Goodbye,' she added, and put down the phone.

Mum came in from the garden and asked Jodie who she'd been talking to.

'It was a lady on the phone,' said Jodie. 'She said Carl would be ready at two o'clock, and asked if we could collect him.' Carl was Jodie's dad's name.

'Oh dear,' said Mum. 'You should have called me. Daddy's meeting must be finishing early. And the car's at the garage, being repaired. We'll have to take a taxi to collect him from the station.'

Mum phoned to book a taxi, and she and Jodie set off just before two o'clock. At the station, they asked the taxi to wait, and went into the big entrance hall. There were lots of people hurrying through the ticket barriers, but there was no sign of Dad. They waited, and waited, but he didn't appear.

'I think I'd better check that the train has arrived,' said Mum. She walked over to the ticket collector.

'Yes,' he told her firmly. 'That train arrived right on time.'

'Dad must have missed it,' Mum told Jodie, 'and there's not another until his usual five o'clock train. We'd better go home.'

They pulled up by the front gate, and got out of the taxi. Mum was just paying the fare, when their own car pulled up in front of them driven by Mr Harris from the garage.

'I was expecting you at two o'clock,' he said to Mum. 'When you didn't arrive, I thought I'd better bring the car round. I knew that you needed it to pick up your husband from the station tonight.'

'Why were you expecting me at two o'clock?' asked Mum. 'You said you would phone to tell me when the car would be ready.'

'But we did!' said Mr. Harris.
Mum turned to Jodie.

'Jodie, what did the lady say on the phone this morning? she asked.

'Oh dear!' said Jodie. 'Maybe I didn't listen properly.'

'No harm done this time,' sighed Mum. 'But I think you and I had better have a little chat about taking messages!'

Jillian Harker

Mr Teazle's toyshop

Mr Teazle was proud of his toyshop. He thought it was very special. But even Mr Teazle did not know just how special. He did not know that after he locked the door and went home each night, something quite amazing happened in his shop. He had no idea that as the clock struck midnight, all his toys came alive!

You can imagine how boring it is to stand on a shelf all day while people stare at you, or prod you. So you will understand how all the toys in the shop looked forward to twelve o'clock when they could jump down from their shelves and have some fun.

One night they were even more excited than usual. There had been a toy delivery, and all the old toys were looking forward to making some new friends.

The fairy doll was the first to run along her shelf after midnight. She had spotted a strange-looking fluffy ball being put on the shelf that afternoon.

'Hello,' she said to the purple ball, 'I'm Flora.'

There was no reply. She tried again – three times – but still the ball said nothing.

'How rude!' said Flora, bending down to the ball. She sighed so loudly that her breath ruffled the purple fluff. The ball giggled.

'I wondered how long it would take you to find out how I work,' said a tiny voice.

'Blow on me again.' Flora blew hard. The ball giggled loudly.

'I think I've found a new friend,' called Sniff, the soft dog, from the shelf below. He nuzzled the grey, plastic puppy that sat next to him.

'Woof!' barked the puppy, as Sniff's nose bumped the black button on its side.

'Press again to hear me sing.' Sniff pressed the button three times. The puppy barked out a loud song.

'I like singing,' it woofed, and started another song straight away.

'Whee! Whoo-oo-oo!' A piercing wail came from a box on the bottom shelf. Flora and Sniff hurried to see what was making the noise. They lifted the box lid. A furry creature with huge eyes stared up at them.

'Whoo-oo-oo! Whee!' yelled the creature – again and again. The grey puppy was still barking. The purple fluffy ball kept giggling. The noise was dreadful.

'Please stop!' begged Sniff after a while, but the new toys took no notice. Other toys in the shop put their hands over their ears. The noise went on.

At last, Flora could stand it no longer. She marched up to the furry creature and slammed the lid of the box shut.

'Me sleep again!' it said softly. All at once, the purple ball sighed and went quiet. The puppy stopped barking. Everything was still.

'What a night!' said Flora. 'I'm exhausted.' She flopped onto the shelf. Sniff nodded. He curled up beside her. They both fell asleep.

When Mr Teazle arrived next morning, he found the fairy doll and the soft dog curled up together on the bottom shelf.

'What on earth are those two doing there?' he wondered. He couldn't begin to imagine. But you know, don't you?

Jillian Harker

Children come from many lands

If I could sail across the sea
I would not find another me.

If I could fly to every place
No other child would have my face.

So many children, everywhere!
Some are dark and some are fair.

Some have brown eyes, green or blue.
Some have skin that's different, too.

For children come from many lands,
islands, mountains, forests, sands.

They may eat fish, chapattis, rice,
noodles, coconuts, food with spice.

Some live in huts with dried mud walls
and some in buildings really tall.

The clothes they wear on certain days
tell us about their people's ways.

Kimonos, ponchos, cool sarong
Sombreros, saris, silk and long.

At festivals on special dates
They each have ways to celebrate.

So many children everywhere
So much to learn, so much to share.

Brenda Williams

Where we live...

Where we live, is a park where I go to play,
and all sorts of shops we can visit each day.

There are baths, where I swim, and jump in the pool,
and a garage where motorists call for their fuel.

There's a post office stores, with a post-box outside,
and a station I know, with trains you can ride.

There's a school and a nursery for children like us,
and bus stops, where sometimes we wait for a bus.

There's a library I go to, with books of all sorts,
right next to a building where people do sports.

There are roads everywhere, and a long underpass,
and telephone boxes, with see-through glass.

There are crossings with lights, and a little green man,
and a market where stallholders unload their van.

There's a hospital here, and a doctor's as well,
And a café, where I like the cakes that they sell.

I like where I live, so I think I shall stay,
I like all the things I can see everyday.

Brenda Williams

Where we live...

Making cakes

On the telephone

On the telephone

Smiling faces

Things to do

Day _____   Date _____

Things to do today   Tick here

Send a fax

To:

From:

Date:

Message:

Page 1

Send a fax

Have you heard them?

Listen to the sounds that these things make.
Tick the boxes when you hear them.

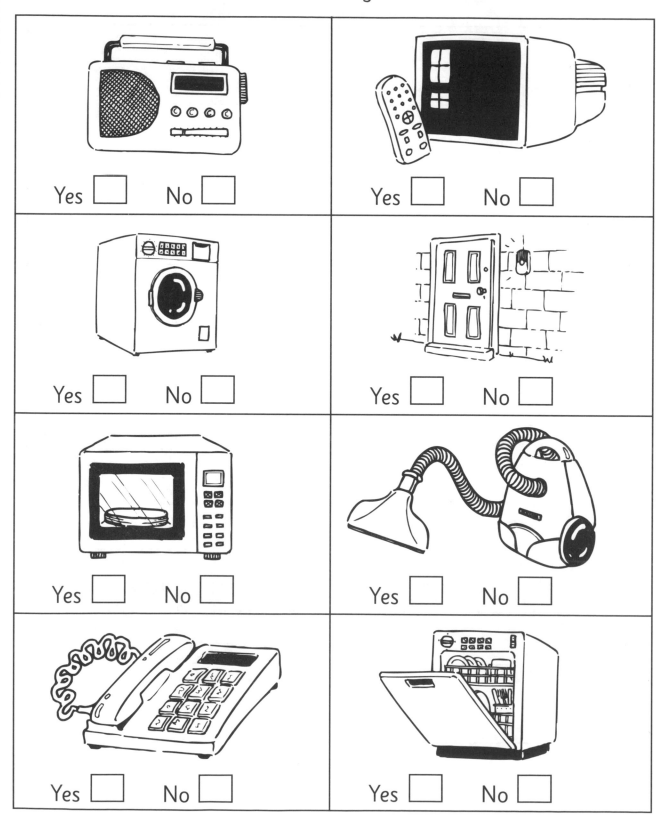

Yes ☐   No ☐

Yes ☐   No ☐

Yes ☐   No ☐

Yes ☐   No ☐

Yes ☐   No ☐

Yes ☐   No ☐

Yes ☐   No ☐

Yes ☐   No ☐

I can make shapes

I can draw a

I can draw a

I can fill my shapes

I can use these colours

I can make shapes smaller

I can make shapes larger

I can make shapes the same size

Keeping it cool

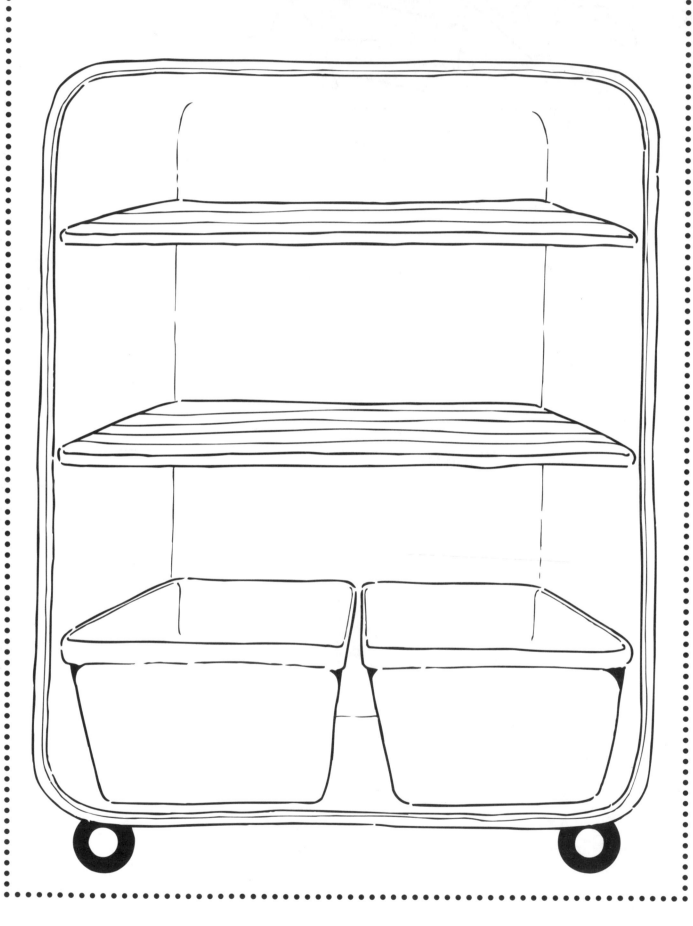

Keeping it cool

Calculator fun

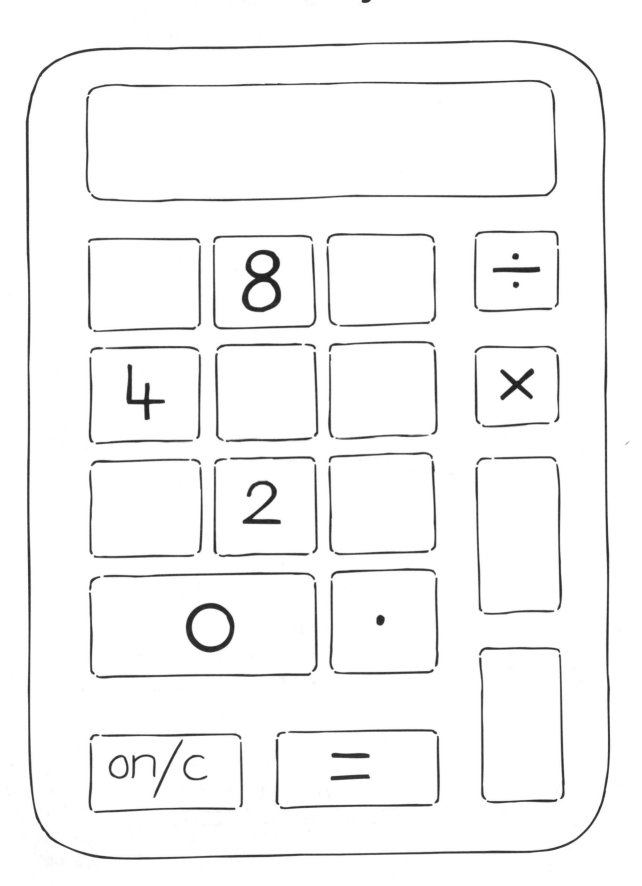

Can you weigh it?

Draw the things that you weighed.
What did the scales show?

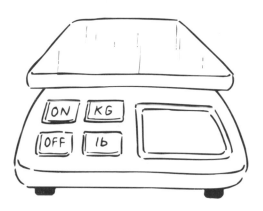 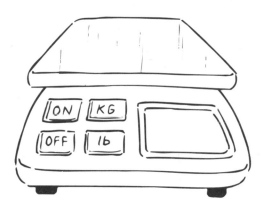

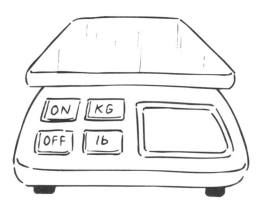

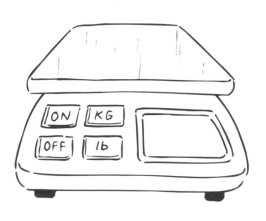 

ICT is all around

Put a tick in the boxes when you find the objects.

The cooker

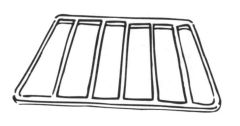

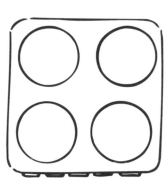

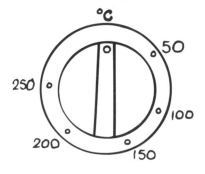

What's missing?

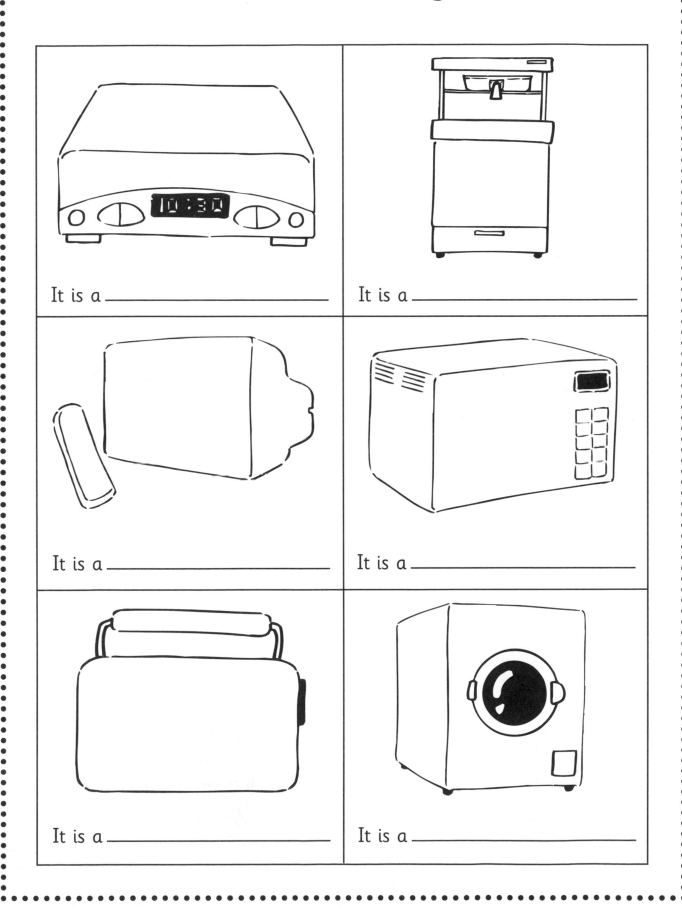

It is a _____

It is a _____

It is a _____

It is a _____

It is a _____

It is a _____

The racetrack game

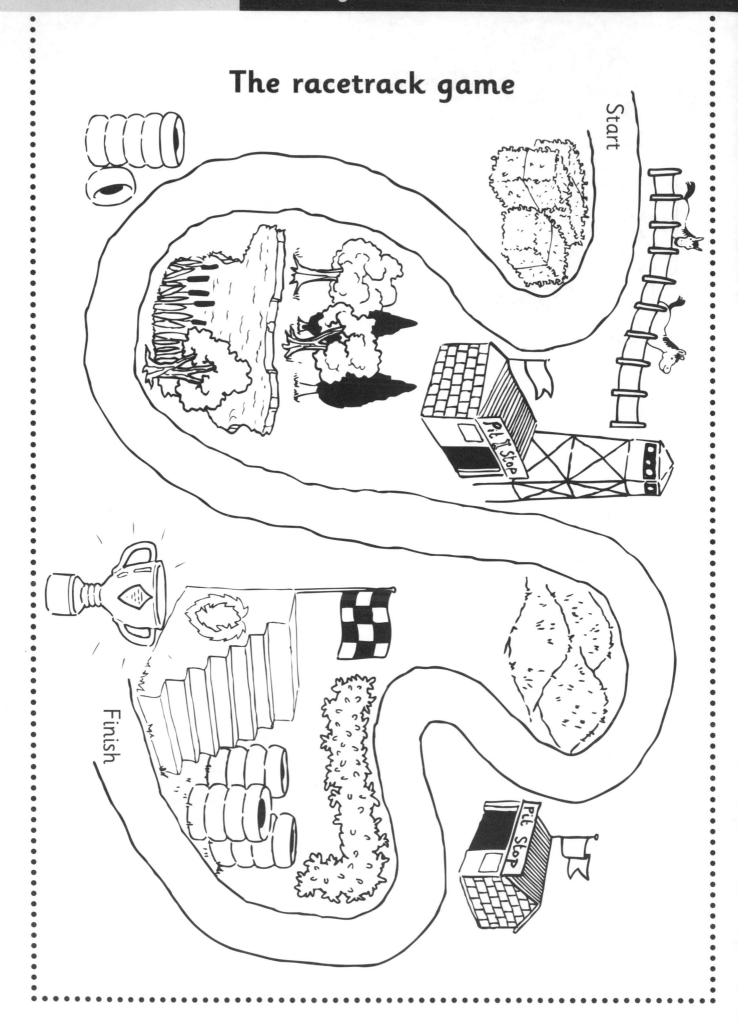

My favourite healthy meal

Draw your favourite food.

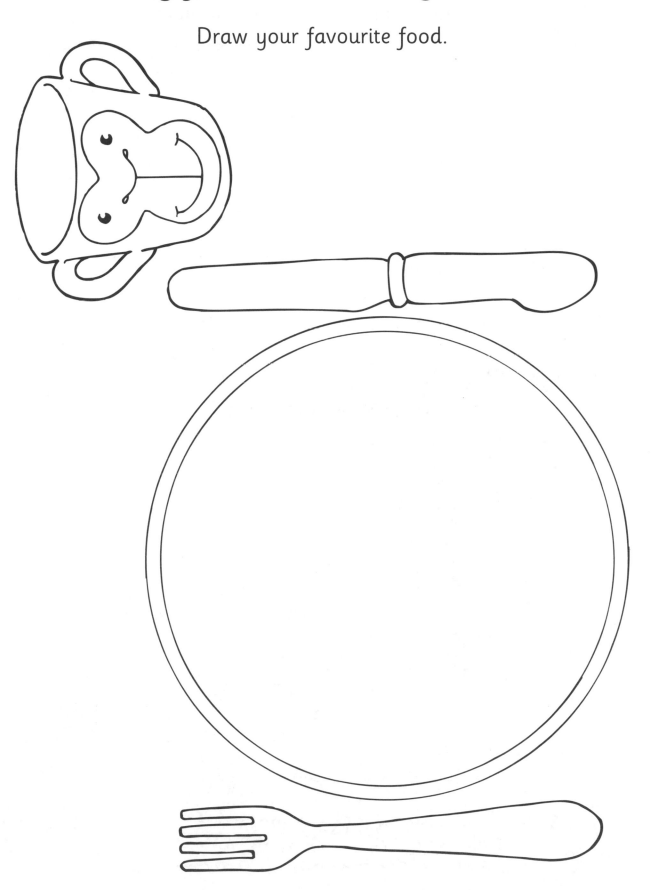

Mouse maze

Help the mouse to find her way home.

EARLY YEARS ACTIVITY CHEST ICT activities

Frame it!

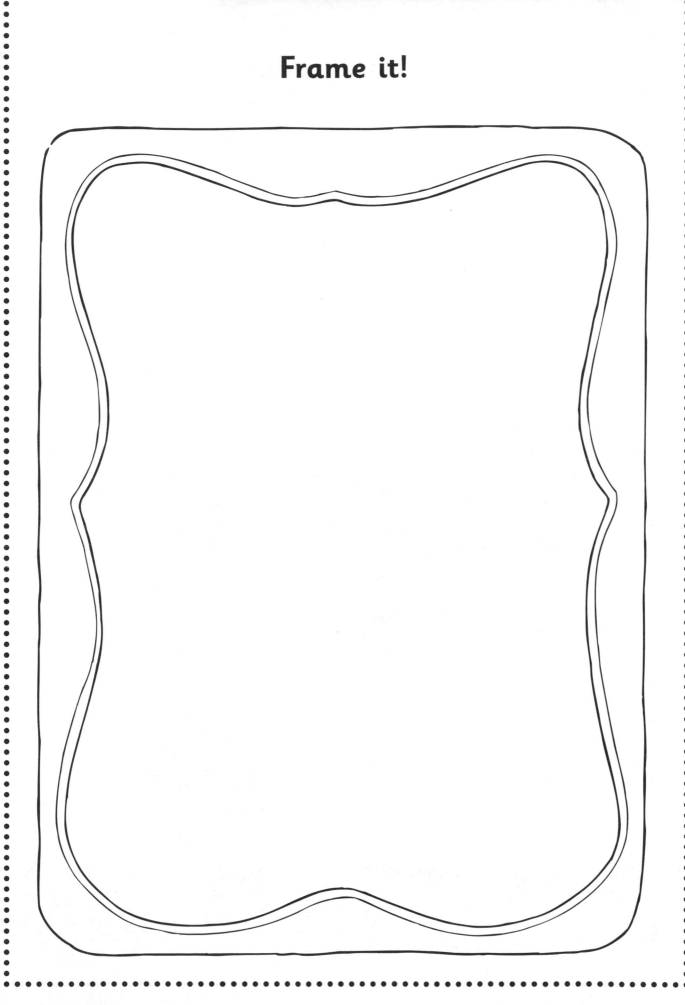

At the garage
Circle the odd ones out.

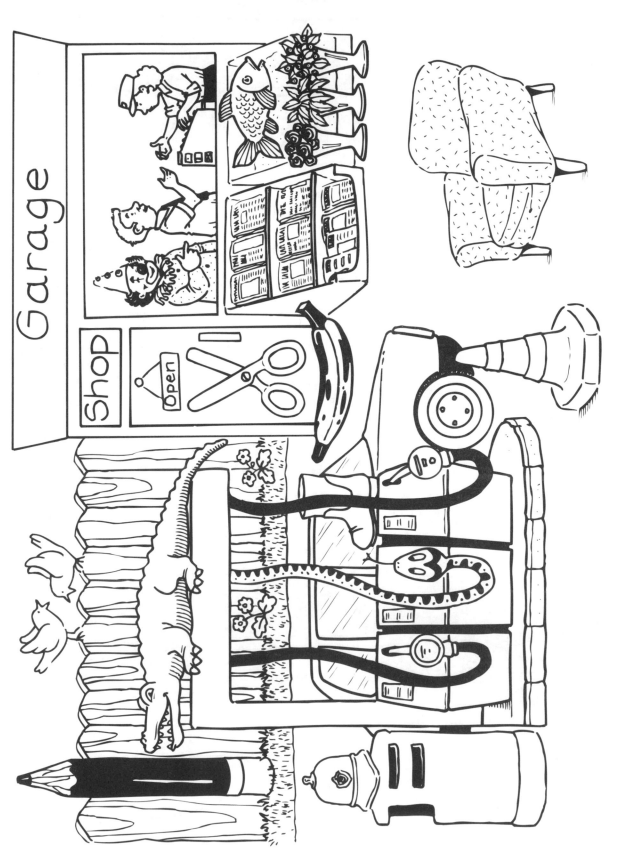

Post it!

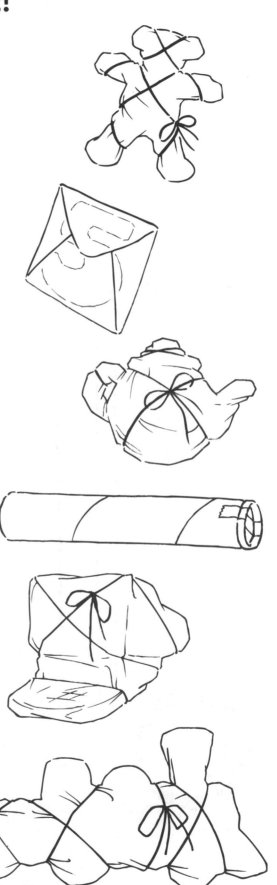

computer

poster

birthday
card

teddy

train

teapot

Post it!

In the trolley

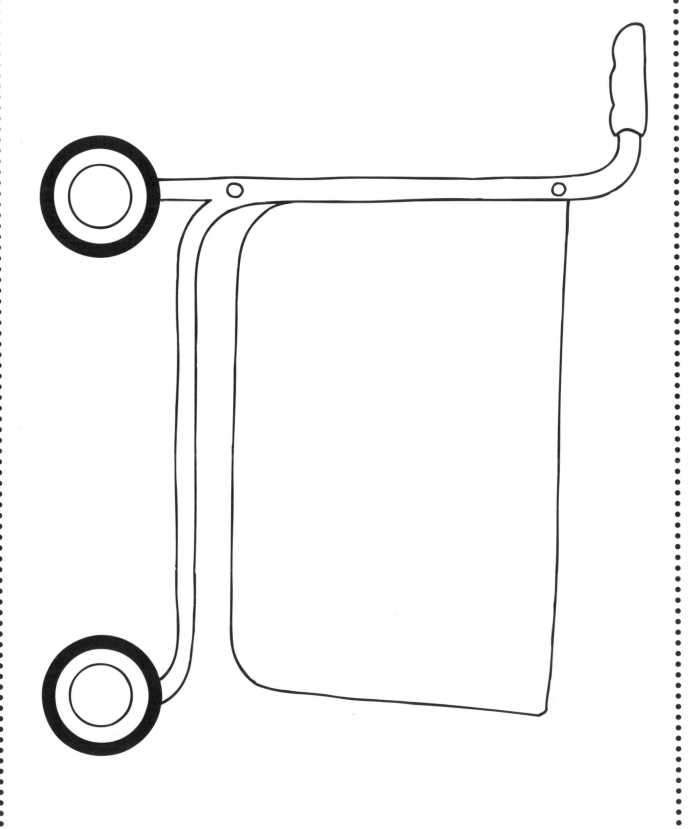

In the trolley